Dear Ravi,

Here is a copy of my recently coauthored book on Asian Immigrants and their family ties. I came to the US as a family sponsored immigrant — Thanks to my brother/sister in law.

I had a swell time this semester at the Economics Department of Emory. Hoping that our paths will cross soon. I am heading to Vermont.

Wishing you the best,

PHANI

Human Capital Investment

"The authors attack forcefully an established myth in the USbased immigration literature, namely that post-1965 immigrants were of low quality since they came at a time with a policy focus on family unification. They make us aware of the huge impact family investments have had among their studied Asian immigrants on high school attendance rates and other human capital improvements generating the most upwardly mobile among American workers. It seems that the migration history of that period has to be re-written."
—Klaus F. Zimmermann, President Global Labor Organization, *Bonn University and UNU-MERIT, Maastricht.*

"Overturning received wisdom on an important policy issue is a rare feat. This book, by leading scholars, manages just that beautifully and convincingly on immigration policy. It will become an instant classic."
—Jagdish Bhagwati, University Professor, *Columbia University*, Author of *In Defense of Globalization* (Oxford)

"*Human Capital Investment* exposes the substantial upward mobility of U.S. immigrants from Asia. Through meticulous data analysis, the authors reveal robust earnings growth for Asian immigrants and refugees by including initial earnings and those who are not working, in school and self-employed. These findings—together with those presented in a new volume of the *Annals of the American Academy of Political and Social Science: Refugee and Immigrant Integration: The Promise of Local Action* (July 2020)—have much to offer to scholars and policy makers seriously concerned about immigrant and refugee integration."
—Katharine M. Donato, Donald G. Herzberg Professor of International Migration, and Director of the Institute for the Study of International Migration, at *Georgetown University's Walsh School of Foreign Service.*

Harriet Duleep • Mark C. Regets
Seth Sanders • Phanindra V. Wunnava

Human Capital Investment

A History of Asian Immigrants and Their Family Ties

Harriet Duleep
Department of Public Policy
William & Mary
Williamsburg, VA, USA

Mark C. Regets
National Foundation for American Policy
Arlington, VA, USA

Seth Sanders
Department of Economics
Cornell University
Ithaca, NY, USA

Phanindra V. Wunnava
Department of Economics
Middlebury College
Middlebury, VT, USA

ISBN 978-3-030-47082-1   ISBN 978-3-030-47083-8 (eBook)
https://doi.org/10.1007/978-3-030-47083-8

© The Editor(s) (if applicable) and The Author(s), under exclusive licence to Springer Nature Switzerland AG 2020

This work is subject to copyright. All rights are solely and exclusively licensed by the Publisher, whether the whole or part of the material is concerned, specifically the rights of translation, reprinting, reuse of illustrations, recitation, broadcasting, reproduction on microfilms or in any other physical way, and transmission or information storage and retrieval, electronic adaptation, computer software, or by similar or dissimilar methodology now known or hereafter developed.

The use of general descriptive names, registered names, trademarks, service marks, etc. in this publication does not imply, even in the absence of a specific statement, that such names are exempt from the relevant protective laws and regulations and therefore free for general use.

The publisher, the authors and the editors are safe to assume that the advice and information in this book are believed to be true and accurate at the date of publication. Neither the publisher nor the authors or the editors give a warranty, expressed or implied, with respect to the material contained herein or for any errors or omissions that may have been made. The publisher remains neutral with regard to jurisdictional claims in published maps and institutional affiliations.

This Palgrave Macmillan imprint is published by the registered company Springer Nature Switzerland AG.

The registered company address is: Gewerbestrasse 11, 6330 Cham, Switzerland

We dedicate this book to
Aarthi and Pooja
Matt, Nick, and David
Jenne, and in memory of Rayleona
In memory of Sri Venkateswararo and Ranganayakamma Wunnava

Foreword

This book criticizes the existing literature on the earnings of immigrants and demonstrates a fundamental flaw in the conventional model. It proposes a richer model that incorporates investments in human capital by immigrants and their families. It challenges the conventional model in three ways:

First, it views the decline in immigrants' entry earnings after 1965 as due to investment in human capital, not to permanently lower "quality."

Second, it adds human capital investment and earnings growth after entry to the model.

Third, by taking investments by family members into account, it challenges the policy recommendation that immigrants should be selected for their occupational qualifications rather than family connections.

It contains a convincing critique of the standard methodology and proposes a better one; that is, analyze each entry cohort separately, include their earnings growth as well as entry earnings, and include only those cohorts that can be followed from entry. The results provide convincing evidence in support of the proposed theory of immigrants' earnings.

The earlier interpretation of the decline in U.S. immigrants' entry earnings since 1965 (as due to a decline in "quality" when Asia and Latin

America replaced Europe as the main source of immigrants) has been controversial. This book should help settle that controversy by providing convincing evidence for a different interpretation.

Professor Emerita of Economics Cordelia W. Reimers
Hunter College and the Graduate School
of the City University of New York
New York, NY, USA
Professor Emeritus of History David M. Reimers
New York University
New York, NY, USA
Author of *Other Immigrants:
The Global Origins of the American People*
New York University Press

Acknowledgements

Support for this book came from many corners, starting with our former professors who introduced us to the theory and measurement of human capital investment: Greg Duncan, Gary Becker, V. Joseph Hotz, James Heckman, Jerry Hausman, and Solomon Polachek. We are also indebted to Barry Chiswick whose influence is evident throughout the book. The conferences he organized and invited us to were key to many of the developments described in this book.

We are especially thankful to Cordelia Reimers for her thoughtful, broad-minded, and painstaking review. Her efforts improved the book in countless ways.

Appreciation is also extended to Margaret Sullivan, the librarian of the former Immigration and Naturalization Service, for helping us to use the INS records that underlie multiple analyses in this study.

Matthew Heisler gave us invaluable help preparing the references, tables, and the index. We would also like to thank Middlebury College Christian A. Johnson Summer Research Fellows, Sarah King, Sung Koo "Kareith" Kim, and Sam Tauke for their research assistance in extracting/cleaning census data and for their invaluable assistance when the details of organizing tables and notes required focused, careful labor.

We would like to thank Palgrave Macmillan Commissioning Editor Elizabeth Graber, editorial assistants Lavanya Devgun and Sophia Siegler and production manager M. Vipinkumar.

We are also grateful to the following individuals who have, in one way or another, supported this effort: Jonathan Dunn, Steve Woodbury, Hal Sider, Stephen Thernstrom, Finis Welch, the late Stanley Lieberson who noted the similarity between our work and his intergenerational work with Mary Waters, and the greatly missed Robert LaLonde.

As documented in this book, the role of family cannot be overemphasized in the journey of immigrants to realize their full potential to be productive and contributing members of our society. One of the authors (Phanindra V. Wunnava) came to the U.S. from India through a *Family Preference* visa. He is immensely grateful to his brother (Professor Subbarao Wunnava) and sister-in-law (Sunanda Wunnava) for sponsoring him to immigrate to the U.S. and giving him adequate timely support to have a chance at the "American Dream." Furthermore, Wunnava would like to thank his spiritual guru, Bhagawan Sri Sathya Baba for giving him strength while working on this book.

Finally, we are indebted to the support, patience, and encouragement of those who have been closest to our excitements and frustrations as we worked on this book, especially Penny Sanders, Matthew, Nicholas, and David Regets, Vijaya Wunnava, Geetha Wunnava, Sanjay Wunnava, and K. Gopalakrishnan Duleep.

Contents

1 Introduction: Background and Overview   1

2 A Brief Review of Immigration from Asia   19

Part I   Theory and Methodology   27

3 What Caused the Decline in Immigrant Entry Earnings Following the Immigration and Nationality Act of 1965?   29

4 The Immigrant Human Capital Investment Model   37

5 Methodological Implications of a Human Capital Investment Perspective   45

Part II   Earnings Growth and Human Capital Investment of Immigrant Men, the 1965–1970 and 1975–1980 Cohorts   53

6 The Earnings Growth of Asian Versus European Immigrants   55

xii CONTENTS

7 The Earnings Profiles of Immigrant Men in Specific Asian Groups: Cross-Sectional Versus Cohort-Based Estimates   65

8 Modeling the Effect of a Factor Associated with Low Entry Earnings: Family Admissions and Immigrant Earnings Profiles   81

9 Human Capital Investment   95

10 Permanence and the Propensity to Invest   107

Part III   A Family Perspective   117

11 Family Income   121

12 Explaining the High Labor Force Participation of Married Women from Asian Developing Countries   135

13 Husbands and Wives: Work Decisions in a Family Investment Model?   151

14 Following Cohorts and Individuals Over Time: Work Decisions of Married Immigrant Women   157

15 Unpaid Family Labor   169

16 Beyond the Immediate Family   177

Part IV   More Recent Cohorts   187

17 Entry Earnings, Earnings Growth, and Human Capital Investment: The 1985–1990 and 1995–2000 Cohorts   189

Part V	The Impact of Refugee Status	199
18	Factors Associated with Refugee Status	201
19	The Earnings and Human Capital Investment of Southeast Asian Refugee Men: The 1975–1980 Cohort	211
20	Married Refugee Women from South East Asia: The 1975–1980 Cohort	223
21	Refugee Entrants from South East Asia, a Decade After the War: The 1985–1990 Cohort	233
Part VI	A Brief Glance Backward and Conclusion	243
22	A Longer Perspective on Initial Conditions and Immigrant Adjustment	245
23	Conclusion	251

Appendix A: Sample Size Information for Year-of-Entry Cohorts at Entry and Ten Years Later by Age and Education Categories	257
Appendix B: Notes on Historical Data in Chap. 22	263
Index	265

ABBREVIATIONS

IHCI Model   Immigrant Human Capital Investment Model
PUMS   Public Use Microdata Sample

List of Figures

Fig. 1.1 National origin composition of legal immigrants. (Source: Authors' creation based on "Table 2. Persons Obtaining Lawful Permanent Resident Status By Region and Selected Country Of Last Residence: Fiscal Years 1820 To 2015" in U.S. Department of Homeland Security, *2015 Yearbook of Immigration Statistics, Office of Immigration Statistics*, December 2016; https://www.dhs.gov/immigration-statistics/yearbook) — 2

Fig. 1.2 Estimates of immigrant earnings growth based on two methods. (Source: Authors' creation) — 8

Fig. 2.1 Immigrant flows within Asian immigration. (Source: Authors' calculations from the Full Count Population Censuses, 1850–1940) — 22

Fig. 3.1 Median 1989 U.S. earnings of men ages 25–54 who immigrated in the years 1985–1990, by country of origin. (Source: Authors' estimates based on the 1990 Census of Population 5% and 1% public-use samples. Notes: The foreign born are defined as persons born outside of the United States excluding those with U.S. parents) — 32

Fig. 3.2 The relationship between gross domestic product (GDP) per adult and U.S. median initial earnings of immigrant men. (Source: Authors' earnings estimates based on 1990 Census of Population 5% and 1% public-use samples. The statistics on GDP per adult as a percentage of U.S. GDP per adult are from Heston and Summers (1991). Notes: Foreign born are defined as persons born outside of the United States excluding those with U.S. parents) — 34

LIST OF FIGURES

Fig. II.1	Earnings growth scenarios. (Source: Authors' creation)	53
Fig. 6.1	Actual and predicted earnings trajectories for Asian immigrants for 1965–1970 entry cohort by age and years of schooling. Foreign-born/U.S.-born median earnings at entry and ten years later. (Source: Authors' creation)	61
Fig. 7.1	Hypothetical earnings profile of Asian and European immigrants. (Source: Authors' creation)	73
Fig. 7.2	The inverse relationship by age and education. (Source: Authors' creation based on 1980 and 1990 Census PUMS data)	76
Fig. 8.1	Fraction of immigrants admitted based on occupational preference by source country. (Source: Authors' creation based on their compilation of statistics from the 1965–1980 annual reports of the Immigration and Naturalization Service)	82
Fig. 8.2	Fraction of immigrants admitted based on occupational preference by source country. (Source: Authors' creation based on their compilation of statistics from the 1965–1980 annual reports of the Immigration and Naturalization Service)	83
Fig. 10.1	Permanence and the potential return to investment. (Source: Authors' creation)	111
Fig. 13.1	Relative propensity of wife working versus husband's potential return to U.S. investment adjusted for expected permanence: 1985–1990 cohort. (Source: Authors' creation)	155
Fig. 14.1	Work status in current period for married women 25–44, with one child, classified by age of child and whether worked the year before. (Source: Authors' creation)	161
Fig. 17.1	Median earnings, the first five years, of immigrant men from the Asian developing countries as a percentage of U.S.-born men: The 1965–1970, 1975–1980, 1985–1990, and 1995–2000 cohorts. (Source: Authors' creation based on 1970, 1980, 1990, and 2000 Census PUMS data)	190
Fig. 17.2	Median earnings, the first five years and ten years later of immigrant men from the Asian developing countries as a percentage of U.S.-born men: the 1975–1980, 1985–1990, and 1995–2000 cohorts. (Source: Authors' creation based on 1980, 1990, 2000, and 2010 Census PUMS data)	193
Fig. 17.3	Percentage of immigrant men from the Asian developing countries attending school by single years of age, compared with US natives and immigrant men from Western Europe, 2000. (Source: Authors' creation based on 2000 census PUMS data)	194

Fig. 17.4	Percentage of immigrant men from the Asian developing countries attending school by single years of age, compared with US natives and immigrant men from Western Europe, 2010. (Source: Authors' creation based on 2010 census PUMS data)	194
Fig. 19.1	Earnings assimilation, for the 1975–1980 cohort, relative to U.S. born, measured with median earnings, for Indochinese refugee groups and the developing country Asian immigrants. (Source: Authors' creation based on 1980, 1990, 2000 census PUMS data)	216
Fig. 19.2	Percentage attending school by single years of age, 1980. (Source: Authors' creation based on 1980 census PUMS data)	219
Fig. 20.1	The wages of Indochinese refugee married women as a percentage of Western European wages (1975–1980 entry cohort, married women) 25–54 in 1980, 35–64 in 1990. (Source: Authors' creation based on 1980 and 1990 census PUMS data)	231

List of Tables

Table 1.1	National origin composition of immigrant flow	5
Table 1.2	Median entry earnings of 25–54-year-old male immigrants, by region, as a percentage of the earnings of US-born men	10
Table 3.1	Median entry earnings in 1989 of immigrant men, aged 25–54, who entered the United States between 1985 and 1990 as a percentage of the earnings of U.S.-born men, by immigrant region of origin	32
Table 6.1	Entry earnings and ten-year real earnings growth rates of age-education cohorts by region of origin (Median earnings, 1989 dollars, deflated by index of weekly wages)	57
Table 6.2	Entry earnings and earnings after ten years, relative to native born of age-education cohorts by region of origin, men 25–54 years old (Median earnings)	59
Table 7.1	Median entry earnings of 25–64-year-old male immigrants, 1989 dollars (percentage of U.S. native born in parentheses)	66
Table 7.2	Entry education level of the 1975–1979 cohort of 25–54-year-old male immigrants (percent)	68
Table 7.3	Relative earnings of U.S. immigrants, evaluated at U.S. native-born mean level of schooling	68
Table 7.4	English language proficiency of 1975–1980 entry cohort of 25–64-year-old male immigrants (percent)	69
Table 7.5	Entry earnings and earnings 10 years later for immigrant men who entered the U.S. in 1975–1980 and were 25–54 in 1980 (median earnings, 1989 dollars) (percentage of US native born in parentheses)	74

Table 7.6	Entry earnings and earnings growth of the 1975–1980 cohort by age and years of schooling, 1980 and 1990 censuses (median earnings, 1989 dollars by PCD)	75
Table 7.7	Entry earnings and earnings 10 years later of the 1975–1980 cohort, by age and years of schooling, relative to the native born, median earnings 1980 and 1990 censuses	77
Table 8.1	Regression of log (earnings) for immigrant men, 25–64 years old, controlling for percentage of country-of-origin/year-of-immigration cohort admitted on the basis of occupational skills (t-statistics in parentheses)	86
Table 8.2	Regression of log (earnings) for immigrant men, 25–64 years old—the effect of percent admitted via occupational skills, controlling for source country and source country human capital interactions (t-statistics in parentheses)	88
Table 8.3	The interactive effect of admission criteria and education on immigrant earnings: regression of log (earnings) for immigrant men, 25–64 years old, all immigrants (t-statistics in parentheses)	90
Table 8.4	Regression of log (earnings) for immigrant men, 25–64 years old, controlling for percentage of country-of-origin/year-of-immigration cohort admitted via occupational skills by education level (t-statistics in parentheses)	91
Table 9.1	Ability to speak English of the 1975–1980 immigrant entry cohort measured in 1980 and 1990	97
Table 9.2	Effect of recent immigration on occupation, immigrant men Ages 25–64	99
Table 9.3	School attendance of the 1975–1980 immigrant entry cohort measure in 1980 and 1990	100
Table 9.4	Educational level of the 1975–1980 immigrant entry cohort measured in 1980 and 1990	102
Table 10.1	Naturalization rates of 1971 cohort of immigrants from Asia, Western Europe, and Canada	108
Table 10.2	Estimated change in cohort size between 1980 and 1990 and entry earnings: 25–54-year-old men in 1980 who immigrated in 1975–1980	109
Table 10.3	Percentage effect of years since migration on annual earnings of foreign-born Japanese men, ages 25–64, 1980 (Benchmark group is foreign-born Japanese men who immigrated before 1950)	113
Table 11.1	Men's earnings and family income by years since migration to the US	122

Table 11.2	Percentage of immigrant families below the poverty line by year of immigration	124
Table 11.3	Family dissolution among Asian and European-Canadian immigrant groups: percentage of ever-married women who are divorced or separated	124
Table 11.4	Poverty rates by years since migration, married couples only	125
Table 11.5	Average family income by years since migration to the U.S., married couples only	126
Table 11.6	Contribution of children to family income expressed as a percentage of total family earnings (married couple families only)	127
Table 11.7	Other relatives per immigrant family by group: number per 1000 families	128
Table 11.8	Contribution of other relatives to family income expressed as a percentage of total family earnings (married couple families only)	129
Table 11.9	Contribution of wives to family income expressed as a percentage of total family earnings (married couple families only)	130
Table 11.10	Earnings of married women and men, 25–64 years old, recent entrants, 1980; year of immigration: 1975–1980	130
Table 11.11	Hours worked and the propensity to work of foreign-born married women, ages 25–64, 1980	131
Table 11.12	Characteristics of Asian and European/Canadian immigrant married women: 25–64 years old, 1980	132
Table 12.1	Effects on labor force participation of child status, general skills, demand conditions, and US specific skills: pooled logit model estimated for married immigrant women, ages 25–64	137
Table 12.2	Logit coefficients on group-specific estimations for married immigrant women, ages 25–64	139
Table 12.3	The effect of Asian origin on the labor force participation of married immigrant women (benchmark group is European/Canadian immigrant women)	141
Table 12.4	Selected coefficients from pooled logit model adjusting for whether married prior to migration and adult relatives in the home: 25–64-year-old married immigrant women, 1980	142
Table 12.5	Effect of having worked full-time in country of origin on labor force participation in America: married immigrant women, Ages 25–64, who immigrated in 1975–1980	145

Table 12.6	The effect of the availability and certainty of income other than the wife's on the labor force participation of married immigrant women	146
Table 12.7	Selected coefficients from Pooled Probit Model estimated for all married immigrant women and immigrant women married to U.S. citizens (benchmark group is European and Canadian married immigrant women)	147
Table 13.1	Percentage impact of years since migration on annual earnings of immigrant men, ages 25–64, by group, 1980	152
Table 13.2	Group effects on the propensity to work of married immigrant women (ages 25–64) who reported migrating to the U.S. in 1985–1990 and each group's average expected return to the husband's investment in U.S.-specific human capital. Immigrant married women from non-English-speaking Western Europe form the reference group (relative ranking in parentheses)	154
Table 14.1	Relationship between time in the United States and the magnitude of the group-specific effects on women's labor force participation; benchmark group is non-English-speaking Western Europe (asymptotic t-statistics in parentheses)	158
Table 14.2	Relationship between time in the United States and the magnitude of the group-specific effects on the hours worked of married immigrant women. Benchmark group is Non-English Speaking Western Europe (t-statistics in parentheses)	164
Table 14.3	Group-specific effects on the wages of married immigrant women during their first five years in the United States, and ten years later, 1980 and 1990 ordinary least squares estimates[a]; Benchmark group is Non-English speaking Western Europe (t-statistics in parentheses)	165
Table 15.1	Number of unpaid family workers per 1000 families, 1980	170
Table 15.2	Percentage of wives working by whether the husband is self-employed, 1980	171
Table 15.3	The marginal effects of husband's self-employment (and other variables) on the propensity of married immigrant women to work in the labor market, 1980 and 1990. (Married couple foreign-born families where both husband and wife are ages 25–64)	172
Table 15.4	Percentage of husbands who are self-employed by year of entry, 1990 (Married couple foreign-born families where both husband and wife are ages 25–64)	173

Table 16.1	Regression of log (earnings) for immigrant men, 25–64 years old, controlling for percentage of country-of-origin/year-of-immigration cohort admitted via the sibling preference (t-test statistics are in parentheses)	179
Table 16.2	Admission criteria and the propensity to be self-employed: Correlation coefficients (p values in parentheses)	182
Table 16.3	Logistic model of immigrant self-employment (asymptotic t-test statistics are in parentheses)	183
Table 17.1	Entry education level of the 1995–2000 and 2005–2010 cohorts of 25–64-year-old male immigrants (percent)	191
Table 17.2	Initial levels of English proficiency for 25–64-year-old male immigrants who entered the United States in 1985–1990 and 1995–2000	192
Table 17.3	Entry earnings and earnings after ten years for immigrant men (Ages 25–54) who entered the US in 1975–1980, 1985–1990, and in 1995–2000 as a percentage of U.S.-born earnings	196
Table 18.1	US geographic distribution of immigrant families who have been in the US five years or less in 1980 and in 1990	202
Table 18.2	US geographic distribution in 1980 and 1990 of Indochinese families who immigrated between 1975 and 1980	203
Table 18.3	Distribution of 25–64-year-old male Indochinese by education level on 1990 census (percent)	205
Table 18.4	Disability rates in 1980 of the 1975–1980 entry cohort: immigrant men, 25–54 years old (percent)	207
Table 19.1	Entry earnings and Earnings after ten years, measured at the median, for men (ages 25–54) who entered the U.S. in 1975–1980 1989 dollars (percentage of U.S. native born earnings in parentheses)	212
Table 19.2	Entry earnings and earnings after ten years, measured at the mean, for men (ages 25–54) who entered the US in 1975–1980 1989 dollars (percentage of US native born earnings in parentheses)	213
Table 19.3	Comparison of entry earnings and ten-year real earnings growth rates by age and education for 1975–1980 Indochinese and comparison immigrant groups, 1989 dollars	214
Table 19.4	Median earnings of the 1975–1980 cohort during the first 10 years and 20 years later (earnings as a percent of median earnings of U.S. native-born men are in parentheses)	215
Table 19.5	English proficiency of the 1975–1980 cohort of immigrant men during the first five years and ten years later	217

Table 19.6	School attendance of the 1975–1980 cohort during first five years and ten years later: immigrant men (percent)	220
Table 20.1	Mean characteristics in 1980 of married immigrant women from South East Asia: 25–54 years old, immigrated between 1975 and 1980	224
Table 20.2	The propensity to work of married immigrant women from South East Asia during their first five years in the U.S. and ten years later: Cohort who immigrated between 1975 and 1980 (asymptotic t-statistics in parentheses)	226
Table 20.3	Hours worked for working married immigrant women from South East Asia during their first five years in the U.S. and ten years later: Cohort who immigrated between 1975 and 1980 (t-statistics in parentheses)	228
Table 20.4	The earnings of married immigrant women from South East Asia at entry and ten years later (in 1990 dollars): Cohort who immigrated between 1975 and 1980 (t-statistics in parentheses)	230
Table 21.1	Initial education levels of the 1975–1980 and 1985–1990 entry cohorts: Immigrant men, 25–54 years old (percent)	235
Table 21.2	Initial English proficiency of the 1975–1980 and 1985–1990 entry cohorts: immigrant men, 25–64 years old (percent)	236
Table 21.3	Initial disability rates of the 1975–1980 and 1985–1990 entry cohorts: Immigrant men, 25–64 years old (percent)	237
Table 21.4	Receipt of public assistance income during the first five years by the 1975–1980 and 1985–1989 entry cohorts: Immigrant men, 25–64 years old (percent)	238
Table 21.5	Median entry earnings, earnings ten years later, and ten-year growth rates for the 1975–1980 and 1985–1990 entry cohorts; Earnings at entry and ten years later are expressed as a percentage of U.S. native-born median earnings	239
Table 21.6	Mean entry earnings, earnings ten years later, and ten-year growth rates for the 1975–1980 and 1985–1990 entry cohorts; Earnings at entry and ten years later are expressed as a percentage of U.S. native-born mean earnings	240
Table 21.7	Comparison of the school attendance during the first five years of the 1975–1980 and 1985–1990 entry cohorts: Immigrant men, 25–64 years old	241
Table 22.1	Percentage laborers of working immigrants (The numbers in parentheses are the total admitted who report an occupation)	246

Table A.1	Unweighted sample sizes for 1965–1970 entry cohort on 1970 (six 1% files) and 1980 (5% and 1% files) census PUMS: Males, high school or less	257
Table A.2	Unweighted sample sizes for 1965–1970 entry cohort on 1970 (six 1% files) and 1980 (5% and 1% files) census PUMS: Males, more than high school	258
Table A.3	Unweighted sample sizes for 1975–1980 entry cohort on 1980 census (5% and 1% files) and 1990 census (5% and 1% files): Males, high school or less	258
Table A.4	Unweighted sample sizes for 1975–1980 entry cohort on 1980 census (5% and 1% files) and 1990 census (5% and 1% files): Males, more than high school	259
Table A.5	Unweighted sample sizes for 1985–1990 entry cohort on 1990 (5% and 1% files) and 2000 census (5% and 1% files): Males, high school or less	259
Table A.6	Unweighted sample sizes for 1985–1990 entry cohort on 1990 (5% and 1% files) and 2000 census (5% and 1% files): Males, more than high school	260
Table A.7	Unweighted sample sizes for 1995–2000 entry cohort on 2000 census (5% and 1% files) and 2010 American community survey: Males	260
Table A.8	Unweighted sample sizes for Southeast Asian refugee men: The 1975–1980 entry cohort, males, high school or less: 1980 census (5% and 1% files) and 1990 census (5% and 1% files)	261
Table A.9	Unweighted sample sizes for Southeast Asian refugee men: The 1975–1980 entry cohort, males, more than high school: 1980 census (5% and 1% files) and 1990 census (5% and 1% files)	261
Table A.10	Unweighted sample sizes for Southeast Asian refugee men: 1985–1990 entry cohort on 1990 (5% and 1% files) and 2000 census (5% and 1% files): Males, high school or less	261
Table A.11	Unweighted sample sizes for Southeast Asian refugee men: 1985–1990 entry cohort on 1990 (5% and 1% files) and 2000 census (5% and 1% files): Males, more than high school	262
Table A.12	Foreign-born women married to foreign-born men, the 1975–80 cohort: Unweighted sample sizes for 1975–1980 entry cohort on 1980 census (5% and 1% files) and 1990 census (5% and 1% files)	262

CHAPTER 1

Introduction: Background and Overview

In 1965, a family-reunification policy for admitting immigrants to the United States replaced a system that chose immigrants based on their national origin. With this change, a 40-year hiatus in Asian immigration ended. Today, over three-quarters of U.S. immigrants originate from Asia and Latin America. Two issues that dominate discussions of U.S. immigration policy are, how have the post-reform immigrants fared in the U.S. economy and how do they contribute to the U.S. economy?

Notable comparisons by economists of the post-reform immigrants with those who entered in 1940–1960—a period when most U.S. immigrants came from Western Europe—conclude that with the family-based immigration reform, the "quality" of U.S. immigrants fell. A precipitous decline in the initial earnings of immigrant men and estimates of low earnings growth support this conclusion. Yet, the estimates of low earnings growth rest on a flawed methodology. By using a less constrained methodology, a different reality emerges for an important component of the post-reform immigrants—Asian immigrants. Though the book's title is *Human Capital Investment: A History of Asian Immigrants and Their Family Ties*, it is more broadly focused on comparing the earnings profiles, human capital investment, and family-aided assimilation of immigrants from economically developing versus economically developed countries.

Asian immigration has received less attention than Hispanic immigration even though as a percentage of U.S. legal immigration it has surpassed Hispanic immigration (Fig. 1.1). We provide basic information on

© The Author(s), under exclusive license to Springer Nature Switzerland AG 2020
H. Duleep et al., *Human Capital Investment*,
https://doi.org/10.1007/978-3-030-47083-8_1

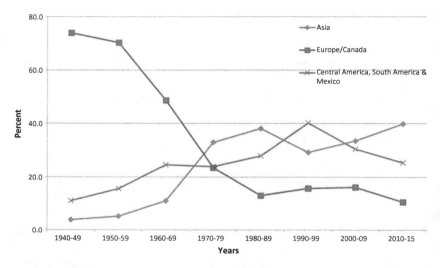

Fig. 1.1 National origin composition of legal immigrants. (Source: Authors' creation based on "Table 2. Persons Obtaining Lawful Permanent Resident Status By Region and Selected Country Of Last Residence: Fiscal Years 1820 To 2015" in U.S. Department of Homeland Security, *2015 Yearbook of Immigration Statistics, Office of Immigration Statistics*, December 2016; https://www.dhs.gov/immigration-statistics/yearbook)

the earnings trajectories and human capital investment of this vital segment of post-reform U.S. immigration.[1] In particular, we:

- test the conclusion of Borjas (1985, 1987, 1992a, b, 1994, 1995, 2015) that due to an increase in the income inequality of the countries contributing to U.S. immigration the quality of U.S. immigrants declined;
- introduce a new model of immigrant human capital investment;
- show how the methodology most economists use understates the earnings growth of immigrants who start with low initial earnings;

---

[1] Several studies offer an historical perspective into the economic characteristics, including educational attainment, of Asian immigrants: Hirschman and Wong 1984, 1986; Model 1988; Barringer et al. 1990; Fernandez and Kim 1998; Nee and Sanders 2001; Sanders et al. 2002; Zhen and Yu 2004.

- explain and illustrate an assumption-free methodology;
- highlight the important connection between immigrant permanence and the propensity to invest in human capital;
- describe the earnings patterns and human capital investment of Asian immigrant men in the groups that dominated the post-1965 reform;
- consider the role of family, particularly women, in human capital investment and earnings patterns;
- describe the economic assimilation of Asian refugee groups, and
- document the intergenerational progress of Asian immigrants who entered the United States before immigration from Asian countries was banned.

HUMAN CAPITAL

Adam Smith described human capital as:

> ... the acquired and useful abilities of all the inhabitants or members of the society. The acquisition of such talents, by the maintenance of the acquirer during his education, study, or apprenticeship, always costs a real expense, which is a capital fixed and realized, as it were, in his person. Those talents, as they make a part of his fortune, so do they likewise of that of the society to which he belongs. (*An Inquiry into the Nature and Causes of the Wealth of Nations*, London: W. Strahan and T. Cadell, 1776.)

In their study of immigrant earnings, economists commonly make three errors: a conceptual error equating earnings with level of human capital, a statistical design error assuming that earnings growth rates are constant across year-of-entry cohorts, and an analytical and policy error of assuming that differences in entry earnings measure differences in immigrant quality. The emphasis on the low entry earnings of the post-1965 immigrants ignores human capital investment and minimizes the depth of immigrant contributions to the U.S. economy. A perspective that permeates our book, echoing an intergenerational theme of sociologists Lieberson and Waters,[2] is that the economic success of immigrants cannot be measured by their initial ability to market their skills.

---

[2] See, for instance, Lieberson (1980), Lieberson and Waters (1988), and Waters and Lieberson (1992).

Earnings growth is a better indicator of the value of the migration both to the immigrants and to their host country. Family-based immigrants and refugees do not enjoy the immediate high-demand for their skills that, by definition, employment-based immigrants do. But, they experience much higher earnings growth. Human capital that is not immediately valued in the host country's labor market *is* useful for learning new skills. Thus, immigrants who do not initially earn on a par with similarly educated natives provide a flexible source of human capital that supports the ever-changing needs of the U.S. economy.

Part I, which includes Chaps. 3, 4 and 5, presents our theory and methodology. Chapter 3 asks, what caused the large post-reform decline in immigrant entry earnings? Building on a theoretical foundation of Chiswick (1978, 1979), Chap. 4 presents the Immigrant Human Capital Investment (IHCI) model. The IHCI model's most important prediction is that the higher incentive to invest in human capital, by immigrants who lack immediately transferable skills, extends beyond U.S.-specific human capital that restores the value of source country human capital (such as English proficiency), to new human capital in general. Informed by a human capital investment perspective, Part I ends with Chap. 5 describing a straightforward methodology to measure the earnings trajectories of immigrants.

THE CHANGING NATURE OF IMMIGRATION, IMMIGRATION POLICY, AND IMMIGRATION RESEARCH

The Immigration and Nationality Act of 1965 was a bellwether of demographic change. In the 40 years before the Immigration Act, visas were allocated based on the national-origin composition of the U.S. population that existed in 1920, a system favoring West European immigration while discriminating against immigrants from Eastern Europe and, with the help of other laws, excluding almost all Asian immigration.

The 1965 Immigration Act swept away the country-specific quotas and replaced them with a system favoring applicants with U.S. family members. Spouses, minor children, and parents of U.S. citizens were admitted without numerical limitations. Of the numerically restricted visas, 80% were reserved for the adult children and siblings of U.S. citizens (as well as their spouses and children), and for the spouses and children of legal permanent resident aliens. The remaining 20% were allocated based on

Table 1.1 National origin composition of immigrant flow

Percent of immigrants originating in

Period	Europe	Asia	Canada	Central America, South America, & Mexico
1940–49	55.2	4.0	18.8	11.2
1950–59	56.2	5.4	14.1	15.7
1960–69	35.3	11.2	13.5	24.6
1970–79	19.4	33.1	4.2	23.9
1980–89	10.7	38.3	2.5	28.0
1990–99	13.8	29.3	2.0	40.3
2000–09	13.9	33.7	2.3	30.6
2010–15	8.8	40.0	1.9	25.5

Source: US Department of Homeland Security, 2015 Yearbook of Immigration Statistics, Office of Immigration Statistics, December 2015. http://www.dhs.gov/immigration-statistics/yearbook

Notes: The numbers in the chart are calculated from "Table 2 Persons Obtaining Lawful Permanent Resident Status by Region and Selected Country of Last Residence: Fiscal Years 1820–2015" in US Department of Homeland Security

Immigrants are defined as lawful permanent residents. Also known as green card holders, these are persons who have been granted permanent residents in the United States

National origin is defined as a country of last residents as opposed to country of birth

occupational skills.[3] This system, which remains largely intact, is the topic of much policy debate.

The new legislation transformed U.S. immigration. With a foothold in the United States, Asian immigration quickly grew. In marked contrast to earlier, European-dominated immigration, recent U.S. immigrants are predominantly Asian and Hispanic (Table 1.1), most come from countries that are less economically developed than the United States, and most are admitted via kinship ties. There has also been a pronounced post-1960 growth in refugee admissions.[4]

---

[3] This taxonomy is approximate and leaves out several categories. For a more comprehensive and detailed description of the various types of U.S. immigrants, refer to Immigration and Naturalization Service (1993). For a longer and broader view of the history of U.S. immigration policy, refer to Bernard (1980), Hutchinson (1981), and Reimers (2005).

[4] Borjas (1990, p. 33) notes, "The fraction of total immigration attributable to refugee admissions increased from 6 to 19% between the 1960s and the 1980s and is rapidly approaching the level reached immediately after World War II (25%), when a large flow of displaced persons entered the United States." The increase in refugee admissions affected the country-of-origin composition of immigrants as well (Reimers 1996). Unlike other legal immigrants, refugees do not need to have U.S. relatives or specific occupational skills to

These trends were historically and prophetically summarized in D. Reimers (1996):

> Historically, most immigrants to the United States have hailed from Europe... As a result of refugee policies, changes in immigration law and amnesties for undocumented aliens, American immigration patterns have undergone major changes in the last 30 years. The numbers have been steadily up since the end of World War II. About 3 million came in the 1960's, 4 in the 1970's, and 6 in the 1980's. Immigration will probably be near 10 million in the 1990's, making it the decade of the largest number of immigrants in American history. And the newcomers...have come from South and East Asia, the Middle East, Mexico, the Caribbean and South America. European immigration accounted for only about 10% of newcomers in the 1980's...Only 5% of immigrants after 1965 came under the occupational categories; the vast majority used the family preference system...The backlog of persons awaiting an American visa in early 1992 was almost 3 million, mostly in Third World countries.... unless Congress decides to revamp immigration policy again, we will see continued domination of immigration from the same nations that prevailed in the 1980's.

Changes in Immigration Research

As immigration changed, so did immigration research. The first labor-economics studies, initiated by Chiswick (1978, 1979, 1980), were primarily "cross-sectional studies." Analyzing census data for a single year, immigrant earnings trajectories are estimated by pairing the initial earnings of recent immigrants with the earnings of immigrants who had been in the United States for multiple years. The difference between the two points is an estimate of immigrant earnings' growth.

The cross-sectional strategy for estimating immigrant earnings growth assumes that once observable demographic and human capital characteristics, such as age and years of schooling, are adjusted for, entry earnings as well as earnings growth remain constant across year-of-entry immigrant cohorts. Using the cross-sectional method, Chiswick and others found that U.S. immigrants had low initial earnings but high earnings growth with their earnings growth substantially exceeding that of U.S. natives.

---

qualify for admission. Instead, their admission to the United States is based on the threat of persecution in their country of origin. For a discussion of U.S. refugee policy and a history of relevant legislation, see Bhagwati (1996) and D. Reimers (1996, 2005).

This optimistic picture was shattered when Borjas (1985, 1987, 1992a, b, 1994) revealed a steep decline in the entry earnings of immigrant men that persists controlling for inter-cohort changes in schooling and age. Borjas showed that in contrast to the more recent cohorts, the initial earnings of the pre-1965 cohorts were never low to begin with. Moreover, Borjas estimated immigrant earnings growth that was no greater than that of U.S. natives. Pairing low initial earnings with low earnings growth produced a dismal picture of the post-1965 immigrants.

The dramatic fall in the adjusted entry earnings of immigrants inspired a fresh fleet of studies armed with a new methodology for measuring immigrant earnings growth. The fixed-cohort-effect methodology, pioneered by Borjas (1985) was first estimated with data from two decennial censuses. It is now standard methodology, presented in labor economics textbooks, and used by most economists with various types of data, including longitudinal data on individuals.

As in the cross-sectional approach, the fixed-cohort-effect methodology estimates immigrant earnings growth in an earnings regression by statistically measuring the relationship between years since migration and immigrant earnings, controlling for age and years of schooling. However, it adds to the estimation a categorical (aka zero-one or dummy) variable for each year-of-entry immigrant cohort. The categorical variables allow the estimated relationship between years since migration and earnings to begin at different levels of initial earnings. The analysis implicitly assumes that earnings growth does not change when entry earnings change.

Figure 1.2 illustrates the two methodologies for estimating immigrant earnings growth. The left-hand side displays the cross-sectional method. It shows the earnings that we would observe in a single cross section from census year t. We see the entry earnings of the most recent cohort (Point A) and the earnings that the earlier cohort (cohort t-10) achieves after ten years in the United States (Point D). Unobserved, at time t, are the earnings that the earlier cohort of immigrants first started with when they came to the United States (Point C). Pairing the initial earnings of the recent cohort (cohort t) with the earnings at the ten-year point of the earlier cohort (cohort t-10) overestimates the earnings growth of the earlier cohort. Nevertheless, the line A-D will accurately represent the earnings trajectory of the more recent cohort *if* its earnings growth exceeds that of the earlier cohort to such an extent that the earnings of the recent cohort catch up to the earlier cohort's earnings at the ten-year mark.

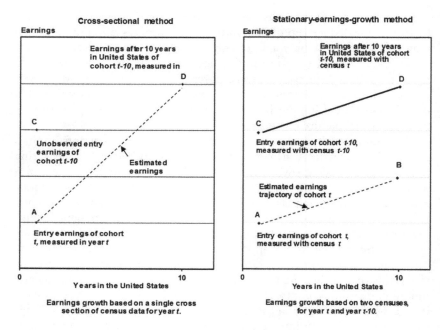

Fig. 1.2 Estimates of immigrant earnings growth based on two methods. (Source: Authors' creation)

The right-hand side of Fig. 1.2 displays the fixed-cohort-effect methodology. It shows the earnings that we would observe by pooling data from two decennial censuses, one from census year t, the other from census year t-10. With the addition of the earlier data, we now observe the initial earnings of cohort t-10 (Point C). The line C-D is the actual earnings trajectory of this earlier cohort. The line A-B is the projected earnings trajectory of the more recent cohort (cohort t). It will accurately predict the more recent cohort's earnings trajectory *if and only if* there has been no inter-cohort change in immigrant earnings growth.

Borjas correctly proved that in a situation where immigrant initial earnings are falling over time, the cross-sectional methodology (pairing the initial earnings of more recent immigrants with the earnings achieved by earlier immigrants after 10 to 15 years in the country) overstates the earnings growth of the *earlier* immigrants. This does not imply, however, that the earnings growth of earlier cohorts predicts the earnings growth of more recent cohorts.[5]

THE TIME PERIODS AND GROUPS THAT ARE STUDIED

A study of Asian immigrants speaks to many of the concerns about the economic desirability of the post-1965 immigrants. In contrast to Western Europe and Canada (countries that dominated pre-1965 U.S. immigration), the per capita GNP of Asian countries (other than Japan) is below that of the United States. Whereas European income inequality is generally less than that of the United States, the reverse is true of Asian countries such as the Philippines and India. Recent immigrants from Asia (with the exception of Japan) have been predominantly admitted via kinship ties; European and Canadian immigrants are more likely to be admitted via employment visas. Asia has also been a major source of refugees. Finally, the initial U.S. earnings of Asian immigrant men, with the exception of Japanese immigrants, lie far below those of immigrant men from Western Europe and Canada (Table 1.2).

Parts II, III, and IV examine the earnings and human capital investment of working-age foreign-born men and women in the groups that dominate post-1965 (non-refugee) Asian immigration in comparison with the groups that dominated pre-1965 immigration—immigrants from Western Europe and Canada.[6] Ranked by their 1980s numerical importance to U.S. immigration, the Asian groups include immigrants from the Philippines, China, Korea, India, and Japan. We pay particular attention to those immigrants who entered the United States from 1965 to 1980, when the entry earnings of immigrant men fell so steeply. Part II examines the earnings growth and human capital investment of immigrant men who came to the United States in 1965–1970 and in 1975–1980.

---

[5] The alarm bells about the declining quality of U.S. immigrants sounded when the entry earnings of post-1965 immigrants were compared with the entry earnings of immigrants who entered in the years 1940–1965. Yet, this benchmark period is one in which immigration was severely restricted. When people think of the immigrants who "built America," they are generally thinking of the immigrants who came in before immigration was restricted. Had recent immigrants been compared to those who entered in the previous great wave of immigration, the alarm bells about declining immigrant quality might not have been set off.

[6] Refer to Model (1988) for a review of literature on the determinants of socioeconomic success among immigrants from Europe and Asia.

Table 1.2 Median entry earnings of 25–54-year-old male immigrants, by region, as a percentage of the earnings of US-born men

	1965–1970 cohort, 1969 earnings	1975–1980 cohort, 1979 earnings	1985–1990 cohort, 1989 earnings
Asia	59%	50	44
Western Europe	83	90	81
Former Communist Europe	83	52	63
Canada	108	116	109

Source: Authors' estimates based on a 1970 Census of Population 6% microdata sample created by combining six 1% Public-Use Sample files, a 6% 1980 file created by combining and reweighting the 5% and 1% files of the 1980 Census of Population, and a 6% microdata 1990 sample created by combining and reweighting the 1990 Census of Population Public Use 5% and 1% Public-Use samples. Appendix A provides information on sample sizes by entry cohort

Notes: For the three 1970 1% Census Files with no year-of-immigration data, residing abroad 5 years before is used as an indicator of being in the 1965–1970 cohort. Native born are defined as persons born in the U.S.; foreign born are defined as persons born outside of the U.S. excluding those with U.S. parents. The sample does not exclude students, the self-employed, and persons with zero earnings

Sullivan (1994) argues that immigration policy has largely ignored the role of women. Resonating with her critique, a consensus has grown that understanding the strategies that immigrants use to adapt to their new countries requires a family perspective. Moreover, men's labor market outcomes may not be fully understood without considering the activities of their wives (Jensen 1991; Morrison and Lichter 1988).

Foner (1997, 1998), Donato (1993), Donato and Gabaccia (2015), Gurak and Kritz (1992), Kahn and Whittington (1996), Morokvasic (1984), Reimers (1985), Tienda (1980), and Zlotnik (1995) are among those who highlight the role of women in facilitating economic assimilation.[7,8] Jensen (1991) found that the ameliorative impact of secondary earners is generally greater for immigrant than native families. Stier

---

[7] A considerable body of research has focused on the role that women play within the household re decisions about work and how working affects the relationship with their husbands e.g., Espiritu (1997), Gilbertson and Gurak (1993), Grasmuck and Pessar (1991), Hondagneu-Sotelo (1992, 1994), Kibria (1993), Landsdale and Ogena (1995), Lim (1997), Pessar (1987), and Pesquera (1993).

[8] Also key to understanding the role of women in economic assimilation are studies of immigrant fertility, alone, and linked to women's labor force participation (e.g., Bean et al. (2000), Blau (1992), Bloom and Killingsworth (1985), Duleep and Sanders (1994), Ford (1990), Kahn (1988, 1994), and Swicegood, Bean, Stephen and Opitz (1988)) as well as

(1991) examined six groups of Asian married women and found, consistent with existing theories and research on the economic behavior of nonimmigrant wives, that immigrant women make decisions on whether or not to join the labor force based on the presence of children and expected labor market productivity.[9] Several studies indicate that the labor market behavior of women immigrants varies across ethnic groups (Bean et al. 1985; Perez 1986; Portes and Bach 1985; Stier and Tienda 1992; Boyd 1984; Reimers 1985).

Building on these and other efforts, Part III brings women into the picture. By focusing on the same year-of-entry cohorts as in Part II we address a key question: Does the labor market behavior of immigrant women offset in any way the steep decline in the initial earnings of immigrant men that occurred after the 1965 Immigration and Nationality Act?

Part IV explores whether the patterns revealed in Part II for immigrant men persist in cohorts that are more recent. Although we do not test the family-based analyses of Part III with more recent data, we hope that researchers will use our efforts as beginning blueprints to analyze more recent cohorts.

The Asian groups studied in Parts II through IV include few refugees. With the end of the Vietnam War, Asian immigration changed abruptly. Though Filipinos, Chinese, Koreans, and Indians continued to dominate, a new wave of Asian immigration emerged, distinguished by its region of origin (South East Asia) and nearly universal refugee status. A key issue within the field of international migration is whether refugees fundamentally differ from immigrants (Hein 1995). Part V explores the relevance of the IHCI Model and family-based strategies to this third wave of (often) poorly educated Asian refugees.[10]

The book ends by traveling back in time to the first wave of Asian immigrants—those who entered the United States before the Immigration Act of 1924 effectively barred their immigration. In a nod to the work of

---

studies of immigrant family formation and dissolution ((e.g., Jasso, Massey, Rosenzweig, and Smith (2000), Landsdale and Ogena (1995), Liang and Ito (1999), and Ortiz (1996))

[9] A gendered understanding of immigrant economic assimilation, as opposed to treating gender as a variable arose in the works of Chiswick (1980), Pedraza (1991), Pessar (1999), Powers and Seltzer (1998), and Weinberg (1992). Bean and Tienda (1987) showed that among Hispanics, labor force participation effects of education and English proficiency were stronger for women than men.

[10] Please refer to Hein (1995).

Lieberson and Waters, we compare the socioeconomic status of these early Asian immigrants with their U.S.-born descendants.

Following in the footsteps of the census-based analyses of Chiswick (1978, 1979, 1980) and Jasso and Rosenzweig (1990), we analyze data on individuals from the public-use samples of the 1960 through 2000 U.S. decennial censuses as well as the Census American Community Service files through 2010. Our work illustrates the strengths of the census data, ways to circumvent weaknesses, and approaches to more fully use the data.

Past studies of Asian immigrants often lump all Asians together or create multi-country groups: immigrants from Japan, a highly economically developed country, are combined with immigrants from countries such as the Philippines; refugee groups are combined with non-refugee groups; the foreign born in a particular group are combined with that group's U.S. born. Building upon prior efforts, we study specific immigrant groups by following cohorts defined by their year of entry into America.[11]

In the chapters that follow, differences and similarities across immigrant groups, within groups, and across year-of-entry cohorts test the Immigrant Human Capital Investment model and highlight family participation in immigrant economic assimilation. By emphasizing earnings growth, human capital investment, and families, a nuanced and optimistic picture of the post-1965 U.S. immigration emerges.

References

Bhagwati, Jagdish. 1996. "U.S. Immigration Policy", in *Immigrants and Immigration Policy: Individual Skills, Family Ties, and Group Identities*, H Duleep and PV Wunnava, eds. Greenwich, Conn.: JAI Press.

Barringer H, Gardner RW, Levin MJ. 1985. *Asians and Pacific Islanders in the United States*. New York: Russell Sage Foundation.

Barringer H, Takeuchi DT, Xenos P. 1990. "Education, Occupational Prestige, and Income of Asian Americans". *Sociology of Education* 63 (1): 27–43.

---

[11] For instance, Barringer et al. (1985) analyze specific Asian groups using the 1980 census but do not separately analyze the foreign born within these groups. Exceptions include the studies of Schoeni (1997, 1998) and Cortes (2004). Further, note that there are important divisions ignored in this book that future researchers will want to investigate. For instance, in our analyses of the 1965–79 and 1975–80 cohorts we combine entrants from Taiwan, Hong Kong, and China. A more disaggregated glimpse of "Chinese immigrants" is given in Chap. 17.

Bean, F. Swicegood CG and King AG. 1985. "Role Incompatibility and the Relationships between Fertility and Labor Supply among Hispanic Women", in *Hispanics and the U.S. Economy*, GJ Borjas, M Tienda, eds. New York: Academic Press.

Bean FD, Swicegood GC, Berg R. 2000. "Mexican Origin Fertility: New Patterns and Interpretations," *Social Science Quarterly* 81: 404–420.

Bean FD and Tienda M. 1987. *The Hispanic Population of the United States*. New York: Russell Sage.

Bernard, William S. (1980) "Immigration: History of U.S. Policy," in Stephan Thernstrom, ed. *Harvard Encyclopedia of American Ethnic Groups*. Harvard University Press, Cambridge, MA and London, England, pp. 486–495.

Blau FD. 1992. "The Fertility of Immigrant Women: Evidence from High Fertility Source Countries," in *Immigration and the Work Force: Economic consequences for the United States and source areas*, GJ Borjas, RB Freeman, eds. Chicago: University of Chicago Press.

Bloom, David and Killingsworth, Mark, (1985), "Correcting for truncation bias caused by a latent truncation variable," *Journal of Econometrics*, 27, issue 1, p. 131–135.

Borjas GJ. 1985. "Assimilation, Changes in Cohort Quality, and the Earnings of Immigrants," *Journal of Labor Economics* 3: 463–489.

Borjas GJ. 1987. "Self Selection and Immigrants," *American Economic Review* 77: 531–553.

Borjas, George J. *Friends or Strangers: the Impact of Immigrants on the U.S. Economy*. New York: Basic Books, 1990.

Borjas GJ. 1992a. "National Origin and the Skills of Immigrants," in *Immigration and the Work Force*, GJ Borjas, RB Freeman, eds. Chicago: The University of Chicago Press.

Borjas GJ. 1992b. "Immigration Research in the 1980s: A Turbulent Decade," in *Research Frontiers in Industrial Relations and Human Resources*, D Lewin, OS Mitchell, P Sherer, eds. Ithaca, N.Y.: Industrial Relations Research Association.

Borjas GJ. 1994. "The Economics of Immigration," *Journal of Economic Literature* 32(4): 1667–1717.

Borjas 1995, "Assimilation and Changes in Cohort Quality Revisited: What Happened to Immigrant Earnings in the 1980s?" *Journal of Labor Economics*, 13(2), 201–245.

Borjas 2015, "The Slowdown in the Economic Assimilation of Immigrants: Aging and Cohort Effects Revisited Again," *Journal of Human Capital*, 9(4), 483–517.

Boyd, Monica. 1984. "At a Disadvantage: The Occupational Attainments of Foreign Born Women in Canada," *International Migration Review* 18(4): 1091–1119.

Chiswick, B.R. (1978) "The Effect of Americanization on the Earnings of Foreign Born Men," *Journal of Political Economy*, October, pp. 897–922.

Chiswick, B.R. (1979) "The Economic Progress of Immigrants: Some Apparently Universal Patterns," in W. Fellner, ed., *Contemporary Economic Problems.* Washington, D.C.: American Enterprise Institute, pp. 359–99.

Chiswick, B.R. (1980) *An Analysis of the Economic Progress and Impact of Immigrants,* Department of Labor monograph, N.T.I.S. No. PB80-200454. Washington, DC: National Technical Information Service.

Cortes, K. E. (2004) "Are Refugees Different from Economic Immigrants? Some Empirical Evidence on the Heterogeneity of Immigrant Groups in the United States," *Review of Economics and Statistics,* 2004, 86 (2), 465–480

Donato, K.P. (1993). "Current Trends and Patterns of Female Migration: Evidence from Mexico," International Migration Research, 27(4):748–771.

Donato, K.P. and Donna Gabaccia, 2015. *Gender and International Migration, from the Slavery Era to the Global Age,* New York: Russell Sage Foundation.

Duleep, H. and Sanders, S. (1994) "Empirical Regularities across Cultures: The Effect of Children on Women's Work," *Journal of Human Resources,* Spring 1994, pp. 328–347.

Espiritu, Yen Le (1997). *Asian American Women and Men.* Thousand Oaks, CA: Sage.

Fernandez M, Kim KC. 1998. "Self-Employment Rates of Asian Immigrant Groups: An Analysis for Intragroup and Intergroup Differences." *International Migration Review* 32(3): 654–681.

Foner N. 1998. "Benefits and Burdens: Immigrant Women and Work in New York City." *Gender Issues* 16(4).

Foner N. 1997. "The Immigrant Family: Cultural legacies and Cultural Changes." *International Migration Review* 31(4): 961–974.

Ford K. 1990. "Duration of Residence in the United States and the Fertility of U.S. Immigrants." *International Migration Review* 24: 34–68.

Gilbertson, G.A. and Gurak, D.T. (1993) "Broadening the Enclave Debate: The Labor Market Experiences of Dominican and Colombian Men in New York City." *Sociological Forum* 8(2):205–20.

Grasmuck S, Pessar P. 1991. *Two Islands: Dominican International Migration.* Berkeley, Calif.: University of California Press.

Gurak, D. and Kritz, M.M. (1992). "Social Context, Household Composition and Employment among Dominican and Columbian Women in New York." Revision of paper published in the Proceedings of the Peopling of the Americas Conference, May.

Hein J. 1995. *From Vietnam, Laos, and Cambodia: A Refugee Experience in the United States.* New York: Twayne Publishers.

Hirschman C, Wong MG. 1984. "Socioeconomic Gains of Asian Americans, Blacks, and Hispanics: 1960–1976," *The American Journal of Sociology,* 90(3):584–607.

Hirschman C, Wong MG. 1986. "The Extraordinary Educational Attainment of Asian-Americans: A Search for Historical Evidence and Explanations." *Social Forces* 65 (1): 1–27.

Hondagneu-Sotelo, P. (1992). "Overcoming Patriarchal Constraints: The Reconstruction of Gender Relations among Mexican Immigrant Women and Men." *Gender and Society*, 6, 393–415.

Hondagneu-Sotelo P. 1994. *Gendered Transitions: Mexican Experiences of Immigration.* Berkeley, Calif.: University of California Press.

Hutchinson EP. 1981. *Legislative History of American Immigration Policy: 1768–1965.* Philadelphia: University of Pennsylvania Press.

Jasso, Guillermina, Massey, Douglas S., Rosenzweig, Mark R., Smith, James P. (2000) "The New Immigrant Survey Pilot (NIS-P): Overview and new findings about U.S. legal immigrants at admission." *Demography.* Feb 37, (1), 127–136.

Jasso G, Rosenzweig MR. 1990. *The New Chosen People: Immigrants in the United States.* New York: Russell Sage Foundation.

Jensen L. 1991. "Secondary Earner Strategies and Family Poverty: Immigrant-Native Differentials, 1960–1980." *International Migration Review* 25(1): 113–140.

Kahn JR. 1994. "Immigrant and Native Fertility during the 1980s: Adaptation and Expectations for Future." *International Migration Review* 28(3): 501–519.

Kahn JR. 1988. "Immigrant Selectivity and Fertility Adaptation in the United States." *Social Forces* 67: 108–27.

Kahn JR, Whittington LA. 1996. "The Impact of Ethnicity, Immigration, and Family Structure on the Labor Supply of Latinas in the U.S." *Population Research and Policy Review* 15: 45–73.

Kibria N. 1993. *Family Tightrope: The Changing Lives of Vietnamese Americans.* Princeton, N.J.: Princeton University Press.

Landsdale, N. and N. Ogena (1995). "Migration and Union Dissolution among Puerto Rican Women," *International Migration Review*, 29(3):671–692.

Liang Z, Ito N. 1999. "Intermarriage of Asian Americans in the New York City Region: Contemporary Patterns and Future Prospects." *International Migration Review* 33(4): 876–900.

Lieberson S. 1980. *A Piece of the Pie: Blacks and White Immigrants Since 1880.* Berkeley, Calif.: Universtiy of California Press 1980.

Lieberson S, Waters MC. 1988. *From Many Strands: Ethnic and Racial Groups in Contemporary America.* New York: Russell Sage Foundation.

Lim IS. 1997. "Korean Immigrant Women's Challenge to Gender Inequality at Home: The Interplay of Economic Resources, Gender, and Family." *Gender and Society* 11: 31–51.

Model, S. 1988. "The Economic Progress of European and East Asian Americans," *Annual Review of Sociology*, 14: 363–380.

Morokvasic, M. (1984). "Birds Of Passage Are Also Women," *International Migration Review*, 18(4): 886–907.
Morrison, Donna Ruane and Daniel T. Lichter (1988) "Family Migration And Female Employment: The Problem Of Underemployment Among Migrant Married Women." *Journal of Marriage and the Family* 50:161–172.
Nee V, Sanders J. 2001. "Understanding the Diversity of Immigrant Incorporation: A Forms-of-capital Model." *Ethnic and Racial Studies* 24(3): 386–411.
Ortiz, Vilma. "Migration and Marriage among Puerto Rican Women." *International Migration Review* 30, no. 2 (1996): 460–84. Accessed July 9, 2020. https://doi.org/10.2307/2547390.
Pedraza S. 1991. Women and Migration: The Social Consequences of Gender. *Annual Review of Sociology* 17: 303–325.
Perez, Lisandro (1986). "Immigrants Economic Adjustment And Family Organization: The Cuban Successes Re-Examined." *International Migration Review*, 20: 4–20.
Pesquera, B. (1993). "In the Beginning He Wouldn't Even Lift a Spoon: The Division of Household Labor." in *Building With Our Hands: New Directions In Chicana Studies*. Ed. A. De La Torre and B.M. Pesquera. Berkeley: University Of California Press. Pp. 181–195.
Pessar, P. (1987). "The Dominicans: Women In The Household And The Garment Industry," In *New Immigrants In New York*, Edited By Nancy Foner. New York, NY: Columbia University Press.
Pessar P. 1999. "The Role of Gender, Households, and Social Networks in the Migration Process: A Review and Appraisal," in *The Handbook of International Migration: The American Experience*, C Hirschman, P Kasinitz, and J DeWind, eds. New York: Russell Sage Foundation.
Portes A and Bach R. (1985) *Latin Journey: Cuban and Mexican Immigrants in the United States*. Berkeley, Calif.: University of California Press.
Powers, Mary G. and William Seltzer, (1998) "Occupational status and mobility among undocumented immigrants by gender." *International Migration Review* 32 (1): 21–56.
Reimers, Cordelia. 1985. "Cultural Differences in Labor Force Participation among Married Women." *American Economic Review, Papers and Proceedings*, (May): pp. 251–55.
Reimers David M. 1996. "Third World Immigration to the United States," in *Immigrants and Immigration Policy: Individual Skills, Family Ties, and Group Identities*, H Duleep and PV Wunnava, eds. Greenwich, Conn.: JAI Press.
Reimers David M. 2005. *Other Immigrants: The Global Origins of the American People*, NY: NYU Press.
Sanders J, Nee V, Sernau S. 2002. "Asian Immigrants' Reliance on Social Ties in a Multiethnic Labor Market." *Social Forces* 81(1): 281–314.

Schoeni, R. (1997) "New Evidence on the Economic Progress of Foreign-Born Men in the 1970s and 1980s," *Journal of Human Resources*, vol. 32, Fall, pp. 683–740.

Schoeni R. 1998. "Labor Market Outcomes of Immigrant Women in the United States: 1970 to 1990." *International Migration Review* 32(1): 57–78.

Stier H. 1991. "Immigrant Women Go to Work: Analysis of Immigrant Wives' Labor Supply for Six Asian Groups." *Social Science Quarterly* 72(1): 67–82.

Stier H, Tienda M. 1992. "Family Work and Women: The Labor Supply of Hispanic Immigrant Wives." *International Migration Review* 26(4): 1291–1313.

Sullivan TA. 1994. "Women Immigrants, Work, and Family," *National Forum* 74(3): 34–36.

Swicegood G, Bean FD, Stephen EH, Opitz W. 1988. "Language Usage and Fertility in the Mexican-Origin Population of the United States." *Demography* 25(1): 17–33.

Tienda M. 1980. "Familism and Structural Assimilation of Mexican Immigrants in the United States." *International Migration Review* 14(3): 383–408.

Waters MC, Lieberson S. 1992. "Ethnic Differences in Education: Current Patterns and Historical Roots," in *International Perspectives on Education and Society* (Vol. 2), A Yogev, ed. Greenwich, Conn.: JAI Press.

Weinberg SS. 1992. "The Treatment of Women in Immigrant History: A Call for Change," in *Seeking Common Ground: Multidisciplinary Studies of Immigrant Women in the United States*, D Gabaccia, ed. Westport, Conn.: Praeger.

Zhen Z, Yu X. 2004. "Asian-Americans' Earnings Disadvantage Reexamined: The Role of Place of Education." *American Journal of Sociology* 109(5): 1075–1108.

Zlotnik, Hania. "The South-to-North Migration of Women." *The International Migration Review* 29, no. 1 (1995): 229–54.

Smith, Adam. 1776. *An inquiry into the nature and causes of the wealth of nations*. London: Printed for W. Strahan; and T. Cadell.

CHAPTER 2

A Brief Review of Immigration from Asia

This chapter reviews immigration from Asia and legislation that shifted immigrant flows within Asian immigration.[1] The landmark Immigration and Nationality Act of 1965 (also known as the Hart–Celler Act) replaced a national quota system whose roots dated back to the Page Act of 1875, which banned Chinese women from migrating to the United States, and the Chinese Exclusion Act of 1882, which banned all immigration from China.

While there were instances of immigration from Asia throughout American history, Chinese men migrating to seek fortune with the discovery of gold in California in 1849 was the first large movement of Asians to the United States. Between 1849 and 1852, more than 20,000 men from China arrived in the United States.[2] With few finding fortune in gold, they moved to other low-skilled occupations including building railroads, farming, and other forms of mining.

Overwhelmingly, Chinese migrants moved to cities, most notably San Francisco. They entered several occupations including fishing, cigar making, and the garment trade. Increasing racial tensions and union action

---

[1] We discuss only immigration and immigration policy that affected large movements of Asian men and women to Hawaii and the U.S. mainland. For a more complete account of immigration from Asia see David M. Reimers *Other Immigrants: The Global Origins of the American People* (2005).

[2] Betty Lee Sung, *The Story of the Chinese in America* (New York: Collier Books, 1967), 22–23.

© The Author(s), under exclusive license to Springer Nature Switzerland AG 2020
H. Duleep et al., *Human Capital Investment*,
https://doi.org/10.1007/978-3-030-47083-8_2

against them eventually drove them out of these trades. Two economic activities open to Chinese immigrants were restaurants and laundry.[3] Nevertheless, their foray into laundry service faced obstacles. Although ultimately curtailed by a Supreme Court decision, the city of San Francisco imposed limits on Chinese laundries.[4]

In 1852, an independent immigration from China to Hawaii commenced. This migration, stimulated by the need of sugar plantations for low-cost, low-skilled labor, brought many Chinese to Hawaii as contract workers. As the end of the nineteenth century neared, more than 45,000 Chinese migrants had arrived in Hawaii.[5] Like the men who migrated to California, many initially planned to work for a short period in Hawaii and then return to China. With time, however, many left the sugar plantations to seek economic opportunities elsewhere. Some remained in rural areas, working in small industries, including private Chinese-owned sugar mills, and in the production of rice, bananas, coffee, and vegetables. Others moved to Honolulu's Chinatown, where established Chinese merchants, traders, carpenters, shoemakers, and professionals employed the newly released plantation workers.[6] After the rise of the Meiji Dynasty, in 1868, contract workers from Japan also arrived in Hawaii.

The Reciprocity Treaty in 1876 allowed Hawaiian sugar and rice to enter the United States duty free. During this rapid rise in the demand for plantation labor, the Chinese Exclusion Act of 1882 passed. This Act was the culmination of anti-Chinese sentiment promoted by protectionist mine labor unions and the Workingmen's Party of California, which gained state political power in 1878 following a six-year period of high unemployment.[7] Some Chinese immigrants, originally recruited to the United States to work in the gold mines or on the railroads, migrated from the West Coast to the islands of Hawaii.[8] The predominant effect of anti-

---

[3] David M. Reimers, *Other Americans* (New York, New York University Press, 2005), 46–48

[4] See *Yick Wo v. Hopkins*, 118 U.S. 356.

[5] Liu, John M., "Race, Ethnicity, and the Sugar Plantation System: Asian Labor in Hawaii, 1850 to 1900." In *Labor Immigration Under Capitalism: Asian Workers in the United States Before World War II*, by Lucie Cheng and Edna Bonacich, 186–210. Berkeley: University of California Press, 1984, 195.

[6] Eleanor C. Nordyke and Richard K.C. Lee, "The Chinese in Hawai'i: A Historical and Demographic Perspective", *The Hawaiian Journal of History*, vol. 23 (1989), 201.

[7] Stephanie S. Pincetl (10 March 2003). "Transforming California: A Political History of Land Use and Development". JHU Press. 23.

[8] Eleanor C. Nordyke and Richard K.C. Lee, (1989) 201.

Chinese sentiment, however, was to change the country of origin of Asian immigration from China to Japan.

After the Chinese Exclusion Act, tens of thousands of Japanese migrants were recruited to Hawaii and then later to the continental United States. Between 1895 and 1908, 130,000 workers from Japan migrated to Hawaii and the continental United States[9]; 30,000 Japanese migrants were recorded in 1907, the peak year of migration.[10] Japanese workers were recruited by railroads and mining interests in the continental United States. Lured by higher wages, many of those who had been recruited to Hawaiian plantations migrated to California, Washington, and Oregon.[11]

Following in the footsteps of earlier Chinese migrants, Japanese migrants left railroad and mining jobs and entered farming, or migrated to West Coast cities. Farming played a larger role for Japanese migrants than for the Chinese: by 1925, half of Japanese migrants were engaged in farming.[12] The Japanese migrants who moved to cities, including San Francisco, Los Angeles and New York City, often worked in agriculture-related businesses or other businesses that required small amounts of capital investment.[13]

As the number of Japanese in the United States increased, resentment against their success in the farming industry and fears of a "yellow peril" grew into an anti-Japanese movement. The primary purpose of the Japanese and Korean Exclusion League, established in 1905, was to restrict migration from Japan and Korea. Their convincing the San Francisco school board to segregate children of Japanese descent in 1906 was the instrument that achieved that goal.

Japan did not want the United States to pass restrictive legislation as was the case with the Chinese. Instead, Japan offered to cease issuing passports for Japanese citizens wishing to work in the continental United States, effectively eliminating new Japanese immigration to the United States. Since Korea was a protectorate of Japan this action eliminated Korean migration as well. In exchange for this "Gentlemen's Agreement,"

---

[9] Eiichiro Azuma, *Between Two Empires: Race, History, and Transnationalism in Japanese America*. (New York: Oxford University Press, 2005), 29.

[10] Masakazu Iwata, 'The Japanese Immigrants in California Agriculture', *Agricultural History* 36.1 (Jan 1996), 25–37.

[11] David M. Reimers, 52–53.

[12] Ronald Takaki, *Strangers from a Different Shore: A History of Asian Americans* (Boston: Little Brown, 1989), 192.

[13] David M. Reimers, 54.

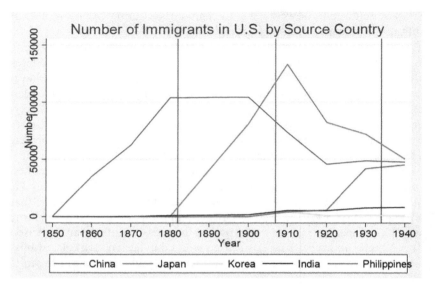

Fig. 2.1 Immigrant flows within Asian immigration. (Source: Authors' calculations from the Full Count Population Censuses, 1850–1940)

the United States agreed to accept the presence of Japanese immigrants already residing in the United States, to permit the immigration of wives, children, and parents, and to avoid legal discrimination against Japanese-American children in California schools.[14]

Figure 2.1 shows the effect of the 1882 Chinese Exclusion Act and the 1907 Gentlemen's Agreement. The first vertical red line marks the 1882 legislation; the second the 1907 agreement. The graph shows that the number of Chinese immigrants stops increasing in 1880 and Japanese migration, which was small before 1880, escalates until 1910 and falls thereafter. We also see a nascent rise in Korean immigration between 1900 and 1910 that reverses after 1910.

Before 1904 there were fewer than one hundred immigrants from India in the United States. In 1907, the number of new Indian migrants

---

[14] C.E. Neu, *An Uncertain Friendship: Theodore Roosevelt and Japan, 1906–1909*. (Cambridge, Massachusetts: Harvard University Press, 1967).

increased to 1072 and continued to increase to 1782 in 1910.[15] However, at no time prior to 1940 did the number of migrants from India exceed 6000.[16]

In the beginning of the twentieth century, several thousand Sikhs migrated to Canada as part of the British Commonwealth after completing their military or police service at British ports in Asia. Rising anti-Indian sentiment in British Columbia led to policy changes and an almost total ban on migrants from India in 1908. Many of the Indians in Canada migrated south, taking up migrant labor positions from the San Joaquin Valley to Sacramento and the Imperial Valley. Others worked on the railroads or lumberyards and mills, the same jobs that migrants from Japan held before the 1907 Gentleman's Agreement.[17] These migrants did not escape anti-Indian sentiment by their move to the United States. The Japanese and Korean Exclusion League was renamed the Asian Exclusion League in 1907 to combat the immigration of Indians as well.[18] The Immigration Act of 1917 passed and included a section of the law that designated an "Asiatic barred zone," which barred migration from much of Asia and the Pacific Islands.

After the Spanish-American War of 1898, the Philippines became a U.S. possession. Filipinos became U.S. nationals, which allowed unrestricted migration to the United States. The Philippines was excluded from the Asiatic barred zone of the 1917 Immigration Act and its successor the Immigration Act of 1924. Just as the limitations on migration from Japan increased migration from India, migration from the Philippines escalated after 1907. The Sugar Planters Association in Hawaii recruited workers from the Philippines.

From 1907 to 1931, when importation of labor temporarily ceased because of the Depression, almost 125,000 Filipinos migrated to Hawaii. Most did not stay in Hawaii but moved to the United States. During the 1920s, the movement from Hawaii to the United States increased and was augmented by a stream of direct migration from the Philippines. By 1930,

---

[15] Chandrasekhar, S. "Indian Immigration in America." *Far Eastern Survey*, vol. 13, no. 15, 1944, 138.
[16] Ibid, 138.
[17] Ibid, 141.
[18] Immigration Policy Center, "The Passage from India, Policy Brief." 2000, 1.

the number of Filipinos recorded by the U.S. Census as residing in California had grown from 2000 to over 30,000.[19]

The Tydings-McDuffie Act in 1934 established the process of the Philippines becoming independent in ten years. It would make migrants from the Philippines aliens for the purpose of immigration. The Act imposed stipulated less than 60 immigrants per year although with the strength of the agricultural lobby this quota was exceeded in most years. Nevertheless, the 1934 Act effectively ended Asia as a source of labor for 30 years.

As our work emphasizes the role of the family in immigrant economic assimilation, it is worth briefly discussing the gender composition of early Asian immigrants. Generally, the hard physical labor on plantations, agriculture, railroads and timber mills was male dominated; the initial migration was overwhelmingly male. As members of each group became permanent in the United States, the issue of marriageable women arose. To a large degree, laws and custom limited intermarriage with white women but some groups, notably Asian Indian and Filipino men, intermarried with Mexican immigrants. The need for marriageable wives typically expressed itself in pressure to change immigration policy to allow more women from Asian home countries to migrate. Groups varied in their success in establishing a family culture; the Japanese were particularly successful as the 1907 Gentlemen's Agreement allowed the wives of immigrant husbands to come to the United States.[20]

The Immigration Act of 1965 dramatically changed the landscape for Asians. The legislation abolished the National Origins system that had systematically limited immigration from Asia since 1924. The Act replaced country-of-origin quotas with preference categories including relatives of U.S. citizens or permanent residents, those with skills useful to the United States, and refugees. Family reunification was a major goal, and the new immigration policy increasingly allowed entire families to migrate together to the United States.

---

[19] Boyd, Monica. "Oriental Immigration: The Experience of the Chinese, Japanese, and Filipino Populations in the United States." *The International Migration Review* 5, no. 1 (1971). 50–51.

[20] David M. Reimers, 2005, op cit., 62–69.

REFERENCES

Sung, Betty Lee, *The Story of the Chinese in America*. New York: Collier Books, 1967.
Boyd, Monica. "Oriental Immigration: The Experience of the Chinese, Japanese, and Filipino Populations in the United States." *The International Migration Review* 5, no. 1 (1971).
Neu, Charles E. *An Uncertain Friendship: Theodore Roosevelt and Japan, 1906–1909*. Cambridge, MA: Harvard University Press, 1967.
Chandrasekhar, S. "Indian Immigration in America." *Far Eastern Surv*ey, vol. 13, no. 15, 1944.
Azuma, Eiichiro, *Between Two Empires: Race, History, and Transnationalism in Japanese America*. New York: Oxford University Press, 2005.
Nordyke, Eleanor C. and Richard K.C. Lee, "The Chinese in Hawai'i: A Historical and Demographic Perspective", *The Hawaiian Journal of History*, vol. 23 (1989).
Immigration Policy Center, "The Passage from India, Policy Brief." 2000.
Cheng, Lucie and Edna Bonacich, *Labor Immigration Under Capitalism: Asian Workers in the United States Before World War II*, 186–210. Berkeley: University of California Press, 1984.
Liu, John M., "Race, Ethnicity, and the Sugar Plantation System: Asian Labor in Hawaii, 1850 to 1900," in Cheng and Bonacich, *Labor Migration under Capitalism*. University of California Press, Berkeley, 1984.
Iwata, Masakazu, "The Japanese Immigrants in California Agriculture", *Agricultural History* 36.1 Jan, 1996.
Reimers, David M. 2005. *Other Immigrants: The Global Origins of the American People*, NY: NYU Press.
Takaki, Ronald, *Strangers from a Different Shore: A History of Asian Americans* (Boston: Little Brown, 1989).
Pincetl, Stephanie S. (10 March 2003). *Transforming California: A Political History of Land Use and Development*. JHU Press.
Yick Wo V. Hopkins, 118 U.S. 356.

PART I

Theory and Methodology

CHAPTER 3

What Caused the Decline in Immigrant Entry Earnings Following the Immigration and Nationality Act of 1965?

In the 40 years before the 1965 Immigration and Nationality Act, the earnings trajectories of immigrants, who were predominantly from Western Europe and Canada, resembled those of U.S. natives of similar age and years of schooling. Their entry earnings were similar, and their earnings growth no greater than that of U.S. natives. After 1965, the initial earnings of immigrants dropped precipitously. The importance of this drop depends on whether it was accompanied by an increase, decrease, or no change in immigrant earnings growth. This chapter describes two hypotheses—with opposing predictions about the relationship between immigrant entry earnings and earnings growth—for the decline in the age- and education-adjusted entry earnings of U.S. immigrants following the 1965 immigration reform.

THE INCOME-DISTRIBUTION/ IMMIGRANT-ABILITY HYPOTHESIS

The Income-Distribution/Immigrant-Ability Hypothesis suggests that the decline in immigrants' initial earnings reflects a decline in immigrant quality resulting from increases in immigration from source countries with high levels of income inequality relative to the United States (Borjas 1987, 1990, 1992a, b, 1994). Immigrants from countries with greater income inequality than the United States are selected from the lower tail of the ability distribution in the sending country; immigrants from countries with less income inequality than the United States are selected from the upper tail of their countries' ability distributions. When countries have relatively egalitarian income distributions,

> ... the source country in effect "taxes" able workers and "insures" the least productive against poor labor market outcomes. This situation obviously generates incentives for the most able to migrate to the U.S. and the immigrant flow is positively selected .... Conversely, if the source country offers relatively high rates of return to skills (which is typically true in countries with substantial income inequality...), the United States now taxes the most able and subsidizes the least productive. Economic conditions in the U.S. relative to those in the country of origin become a magnet for individuals with relatively low earnings capacities, and the immigrant flow is negatively selected. (Borjas 1992b, p. 429)

An empirical test of the income-distribution/immigrant-ability thesis found that the extent of income inequality of source countries is negatively associated with the relative quality of U.S. immigrants, as measured by the initial wage differential between entering immigrants and natives of the same education level (Borjas 1987).[1]

Borjas notes that before the 1965 Immigration and Nationality Act, West Europeans dominated U.S. immigration. The national origins quota system, which admitted persons according to the U.S. ethnic composition of the late nineteenth and twentieth centuries, "encouraged immigration from (some) Western European countries and discouraged immigration from all other countries." Measuring income inequality by the ratio of income accruing to the top versus bottom percentiles of households, Borjas showed that income dispersion in the average immigrant's source

---

[1] A specification error in this empirical test is that the relevant distribution for a potential emigrant, in an analysis that focuses on immigrant earnings controlling for education, is the earnings distribution associated with that person's level of education, not the income distribution of the entire country. The overall earnings distributions of countries may have little relationship to the earnings distributions of individuals with specific levels of education. For instance, a country with a large proportion of illiterates and a large proportion of Ph.D.'s would have an extremely unequal income distribution relative to the overall income distribution of the United States. Yet, the earnings distribution of Ph.D.'s might be narrower in that country than the earnings distribution of American Ph.D.'s. In such a case, it would be the higher quality Ph.D.'s that would have the most to gain by migrating to a country that would reward their skills.

country doubled in the postwar period, with most of that increase occurring after 1960 (Borjas 1992a, p. 44).

The new flow of migrants originates in countries that are much more likely to have greater income inequality than the United States. It would not be surprising, therefore, if the quality of immigrants declined as a result of the 1965 Amendments. (Borjas 1987, 537)

It is theoretically ambiguous whether lower ability leads to lower initial earnings.[2] Under any human capital model, however, a decline in ability would not be associated with an increase in earnings growth. According to the immigrant-ability explanation for the decline in immigrants' initial earnings, immigrant earnings growth should have declined or stayed constant as the entry earnings of immigrants declined. The latter is consistent with the fixed-cohort-effect methodology.

THE ECONOMIC-DEVELOPMENT/ SKILLS-TRANSFERABILITY HYPOTHESIS

An alternative hypothesis for the post-1965 decline in the education- and age-adjusted entry earnings of immigrants is that it reflects a decline in the extent to which the country-of-origin skills of immigrants transfer to the United States.[3] The initial earnings of U.S. immigrants vary enormously depending on where they come from (Fig. 3.1).[4]

Immigrants from the source regions that dominate recent U.S. immigration (Asia and Central and South America) initially earn about half or less than half of what U.S. natives earn, whereas the entry earnings of West European immigrants resemble those of the U.S. born. These differences persist within age and education categories (Table 3.1).[5]

[2] If all factors remain unchanged, higher ability individuals would, theoretically, invest in more human capital than lower ability individuals, which would lower the initial earnings of the higher ability group.
[3] See Duleep and Regets 1997.
[4] Figure 3.1 shows by country of origin the 1989 median initial earnings of working-age immigrant men who entered the United States between 1985 and 1990. The 1989 median earnings estimates for the 1985–90 cohort shown in Fig. 3.1 are based on a 6% microdata sample created by combining and reweighting the 1990 Census of Population 5% and 1% Public Use samples.
[5] Asian immigration is dominated by immigration from less developed countries. In Table 3.1, Asia includes Japan.

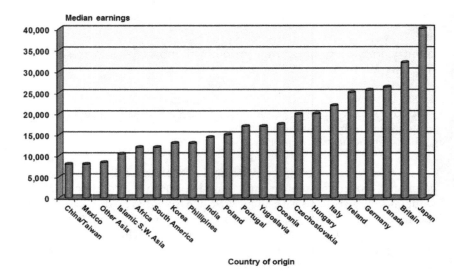

Fig. 3.1 Median 1989 U.S. earnings of men ages 25–54 who immigrated in the years 1985–1990, by country of origin. (Source: Authors' estimates based on the 1990 Census of Population 5% and 1% public-use samples. Notes: The foreign born are defined as persons born outside of the United States excluding those with U.S. parents)

Table 3.1 Median entry earnings in 1989 of immigrant men, aged 25–54, who entered the United States between 1985 and 1990 as a percentage of the earnings of U.S.-born men, by immigrant region of origin

	All	25–39 years old; 1–12 years of school	25–39 years old; more than 12 years of school	40–54 years old; 1–12 years of school	40–54 years old; more than 12 years of school
All immigrants	41%	53	48	38	50
By region of origin					
Asia	44	59	43	32	44
Central/South America	36	51	45	38	40
Western Europe	101	115	93	84	137

Source: Authors' estimates based on a 6% microdata sample created by combining and reweighting the 1990 Census of Population 5% and 1% Public Use samples. Appendix A provides information on sample sizes by entry cohort

Notes: Native born are defined as persons born in the U.S.; foreign born are defined as persons born outside of the United States excluding those with U.S. parents. The sample does not exclude students, the self-employed, and persons with zero earnings

A key factor associated with variation in immigrants' initial U.S. earnings is the source country's level of economic development. Immigrants from regions of the world with levels of economic development similar to the United States, such as Western Europe and Japan, have initial earnings approaching or exceeding those of comparably educated and experienced U.S. natives. Those hailing from economically developing countries have low initial earnings relative to their U.S.-born statistical twins. Plotting the median 1989 U.S. earnings of immigrant men who entered the United States in 1985–1990 against the 1987 per adult GDP of each source country[6] reveals a positive relationship between immigrant entry earnings and level of economic development (Fig. 3.2).[7]

Level of Economic Development and Immigrant Skill Transferability

Borjas noted the increase in the inequality of U.S. immigrant source countries following the 1965 immigration reform. Yet, there was also an

---

[6] The 1987 per adult GDP of each source country is as a percentage of the U.S. per adult GDP. The observations in Fig. 3.2 on U.S. median earnings for immigrant men and GDP per adult as a percentage of U.S. GDP per adult are for the following countries: Argentina, Australia, Bangladesh, Bolivia, Brazil, Canada, Chile, China, Colombia, Costa Rica, Czechoslovakia, Dominican Republic, Ecuador, Egypt, El Salvador, Fiji, France, West Germany, Greece, Guatemala, Guyana, Haiti, Honduras, Hong Kong, Hungary, India, Indonesia, Iran, Ireland, Israel, Italy, Jamaica, Japan, Jordan, The Republic of Korea, Laos, Malaysia, Mexico, Morocco, Myanmar, Netherlands, New Zealand, Nicaragua, Nigeria, Pakistan, Panama, Peru, Philippines, Poland, Portugal, Romania, South Africa, Spain, Sri Lanka, Sweden, Switzerland, Syria, Taiwan, Thailand, Trinidad and Tobago, Turkey, U.S.S.R., United Kingdom, Venezuela, and Yugoslavia. All countries for which we had information on the GDP per adult were included. Median earnings for immigrant men in the 1985–90 cohort from the aforementioned 65 countries were estimated using a 6% microdata sample created by combining and reweighting the 1990 Census of Population 5% and 1% Public Use samples. The statistics on GDP per adult as a percentage of U.S. GDP per adult are from Heston and Summers (1991).

[7] When the median 1989 entry earnings of immigrant men in the 1985–1990 cohort are regressed on source-country GDP, the estimated coefficient indicates that the initial earnings of immigrant men increase 2280 dollars for each 10-percentage-point change in the country-of-origin GDP measure. The R^2 for this regression is .48.

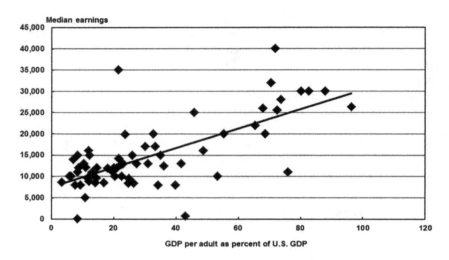

Fig. 3.2 The relationship between gross domestic product (GDP) per adult and U.S. median initial earnings of immigrant men. (Source: Authors' earnings estimates based on 1990 Census of Population 5% and 1% public-use samples. The statistics on GDP per adult as a percentage of U.S. GDP per adult are from Heston and Summers (1991). Notes: Foreign born are defined as persons born outside of the United States excluding those with U.S. parents)

increase in the extent to which immigrants came from less economically developed countries (Reimers 1996).[8]

Two conceptualizations link immigrant skill transferability to the economic development level of immigrant source countries. Chiswick (1978, 1979) proposed that source-country variations in immigrants' initial earnings stem from variations in the skills learned by growing up and working in different source countries. Holding constant immigrants' levels of human capital (years of schooling and work experience), the skills of

---

[8] Borjas (1992a, p.44) also notes: "The changing national origin mix of successive immigrant waves cut by more than half the per capita GNP of the country represented by the typical immigrant, with most of this decline occurring after 1960."

immigrants hailing from economically developed countries transfer more easily to the United States because of similar educational systems, industrial structures, and labor market reward structures; the skills of immigrants from economically developing countries transfer less easily (initially resulting in lower U.S. earnings) as the formal education and work experience in these countries are less applicable to the U.S. labor market.

Certain facts give pause to the conclusion that once we hold years of schooling constant, the skills learned and used in developing countries are less transferable to the United States than the skills learned and used in economically developed countries. International comparisons reveal that mathematical skills acquired at *given* levels of schooling in developing countries can exceed those acquired in U.S. schools (Rivera-Batiz and Sechzer 1991). Moreover, immigrants from some of today's dominant source countries are more proficient in English than immigrants from some West European countries.

An alternative or additional explanation for why U.S. immigrants from economically developing countries are less likely to have highly transferable skills than those from economically developed countries is that people who face constraints in their home countries will be more likely to migrate even if the migration entails substantial human capital investment. Constraints potential migrants face in developing countries include war, gang violence, discrimination, limited professional opportunities, less capitalized research facilities, an inflexible labor market, restrictions on adult education, rigid social structures, and limited opportunities for their children.

From any given country, there would be a mix of those who migrate to the United States and invest in human capital and those who are more transient and do not need to invest in human capital. The proportions would vary, however, across source countries. In economically developed countries, with opportunities and conditions resembling those in the United States, fewer individuals would choose to migrate to the United States when doing so involves substantial investment in human capital.[9]

---

[9] This is not simply a model of migration to the wealthier economy; migrants with highly transferable skills *from either developing or economically developed countries* may face no initial loss in earnings or status.

Regardless of its cause—country-specific institutional differences or the constraint-driven selection mechanism we propose, or both—if variation in skills transferability is the dominant cause of variation in immigrants' initial earnings then the Immigrant Human Capital Investment (IHCI) model (described in the Chap. 4) predicts an inverse relationship between immigrants' education-adjusted entry earnings and their subsequent earnings growth.

REFERENCES

Borjas GJ. 1987. "Self Selection and Immigrants". *American Economic Review* 77: 531–553.

Borjas, George J, 1990. "Self-Selection and the Earnings of Immigrants: Reply," *American Economic Review*, vol. 80(1), pages 305–308, March.

Borjas GJ. 1992a. "National Origin and the Skills of Immigrants," in *Immigration and the Work Force*, GJ Borjas, RB Freeman, eds. Chicago: The University of Chicago Press.

Borjas GJ. 1992b. "Immigration Research in the 1980s: A Turbulent Decade," in *Research Frontiers in Industrial Relations and Human Resources*, D Lewin, OS Mitchell, P Sherer, eds. Ithaca, N.Y.: Industrial Relations Research Association.

Borjas GJ. 1994. "The Economics of Immigration." *Journal of Economic Literature* 32(4): 1667–1717.

Chiswick, B.R. (1978) "The Effect of Americanization on the Earnings of Foreign Born Men," *Journal of Political Economy*, October, pp. 897–922.

Chiswick, B.R. (1979) "The Economic Progress of Immigrants: Some Apparently Universal Patterns," in W. Fellner, ed., *Contemporary Economic Problems*. Washington, D.C.: American Enterprise Institute, pp. 359–99.

Duleep, H. and Regets, M. (1997) "The Decline in Immigrant Entry Earnings: Less Transferable Skills or Lower Ability?" *Quarterly Review of Economics and Finance*, vol. 37, Special Issue on Immigration, pp. 189–208

Heston, Alan and Robert Summers. 1991. "The Penn World Table (Mark 5): An Expanded Set of International Comparisons, 1950–1988," *Quarterly Journal of Economics*, May, pp. 327–368.

Reimers David M. 1996. "Third World Immigration to the United States," in *Immigrants and Immigration Policy: Individual Skills, Family Ties, and Group Identities*, H Duleep and PV Wunnava, eds. Greenwich, Conn.: JAI Press.

Rivera-Batiz, Francisco L. and Sechzer, Selig L. (1991) "Substitution and Complementarity between Immigrant and Native Labor in the United States," in *U.S. Immigration Policy Reform in the 1980s*, eds. F. L. Rivera-Batiz, S.L. Sechzer, and Ira N. Gang, New York: Praeger, pp. 89–116.

CHAPTER 4

The Immigrant Human Capital Investment Model

Theoretically, immigrant skill transferability and the propensity to invest in human capital are inextricably linked. Chiswick (1978, 1979) hypothesized that immigrants entering the United States, or other host country, lack (in varying degrees) the skills specific to that country that would enable their human capital to be fully valued in the U.S. labor market. To bring to life their source-country human capital, immigrants engage in various forms of human-capital investment such as learning English and becoming familiar with the host country's institutions, production methods, and technical terms. As immigrants gain host country-specific skills and credentials, the labor-market value of their source-country human capital increases and their earnings rise.

Duleep and Regets (1999, 2002) articulated an Immigrant Human Capital Investment (IHCI) model that formalizes Chiswick's concept of skill transferability to the host country's labor market and introduces two additional aspects of skill transferability:

(1) Immigrants whose source-country skills do not fully transfer to the host country's labor market will, by virtue of their lower wages, have a lower opportunity cost of human-capital investment than natives or immigrants with high skill transferability: The time they spend learning new skills, instead of applying their current skills to

earning, is less costly than it is for natives or for high-skill transferability immigrants, who earn more with the same level of schooling and experience.

(2) Source-country human capital that is not immediately valued in the host country's labor market *is* useful for learning new skills. Persons who have learned one set of skills—even if those skills are not valued in the destination-country labor market—have advantages in learning new skills: previously learned work and study habits facilitate the learning of new skills. Moreover, common elements between old and new skills aid learning. Cognitive psychologists refer to this phenomenon as "transfer" (Mayer and Wittrock 1996).

To the extent that variations in immigrants' age- and education-adjusted initial earnings reflect variations in initial U.S. skill transferability, the above considerations—linking low skill transferability to high rates of human capital investment, hence high rates of earnings growth—predict an inverse relationship between immigrants' (adjusted) entry earnings and earnings growth. Across groups, the lower the entry earnings, the higher the earnings growth should be[1]; over time, as entry earnings fall, earnings growth should rise, and vice versa.

The Effect of Education and Age on the Propensity to Invest in Human Capital

In most human capital models, the effect of education on the propensity to learn new skills is ambiguous. More education increases the productivity of time spent investing in human capital: with greater schooling, additional learning takes less time and energy, which increases the incentive to invest in human capital. But more education also increases the opportunity cost of time spent in human capital investment, and this reduces the incentive to invest in human capital.

---

[1] Chiswick (1978, 1979, 1980) theorized that because of less-than-perfect international transferability of skills, immigrants initially earn significantly less than the native born with similar levels of education and age. However, immigrants invest in U.S. specific skills, which further lowers their initial earnings but results in high earnings growth with time in the United States. A corollary of Chiswick's hypothesis for understanding earnings differences across immigrant groups is that, "The initial earnings deficiency, and the steepness of the subsequent rise in earnings, would be smaller the greater the similarity between the country of origin and the United States" (Chiswick 1978, p.899).

In the IHCI model, human capital that does not transfer to the labor market increases the productivity of time spent investing in human capital but does not increase the opportunity cost of human capital investment. Thus, the IHCI model predicts that the greater propensity to invest in human capital of low-skill-transferability versus high-skill-transferability immigrants (and of immigrants versus natives) increases with the education level of immigrants.

It is a common result from human capital models and empirical estimates that the young engage in more human capital investment than the old since they have a longer period over which to receive a return from new human capital. In the IHCI model, youth makes investment more likely, and increases the sensitivity of investment to the rate of initial skill transferability. Moreover, longer time horizons increase the likelihood that the more highly educated will have greater rates of human capital investment.

The Importance of Permanence

A key component of the IHCI model is permanence (Duleep and Regets 1999). Individuals who migrate and embark on investments in new human capital would do so only if they could reap the benefits of those investments: *embedded in their decision to migrate is the decision to stay in their adopted country.* Immigrants who do not intend to stay are likely persons who can work without having to invest in new human capital (persons who come via employment-based visas, employees of foreign-owned firms, and persons who come to work in jobs requiring minimal U.S.-specific skills) or for whom host-country human capital investments easily transfer back home. In the absence of programs that recruit workers to meet specific labor market needs, immigrants from economically developing countries would be more likely to have lower U.S. skill transferability and greater permanence than immigrants from economically developed countries.[2]

---

[2] Piracha, Tani, and Vadean (2012) find with the Australian Longitudinal Survey of Immigrants that occupation-education mismatches in the source country predict occupation-education mismatches in the host country. They argue that the mismatch in the source country indicates lower ability, as opposed to skills-transferability issues of those who migrate. Yet, what economists call a skills mismatch in an immigrant's original country, might be more comprehensively defined as a constraint: individuals who choose to migrate may face constraints in their home country that they wish to escape by moving to less restrictive societies

Economic Implications of the IHCI Model

To summarize, the IHCI model rests upon four conceptual building blocks[3]:

(1) The less immigrant skills transfer to the U.S. labor market, the greater the return to investment in U.S.-specific human capital since its acquisition enables immigrants to bring to the U.S. labor market their host-country human capital.
(2) Source-country human capital that does not transfer to the U.S. labor market is valuable in the production of U.S. human capital.
(3) Holding the level of human capital constant, the opportunity costs of human capital investment are less for low-skill-transferability immigrants than for high-skill-transferability immigrants or for natives.
(4) The decision of immigrants to invest in new human capital is jointly determined with their decision to stay in their adopted country.

These concepts imply that among comparably educated individuals, immigrants who intend on staying in the United States will invest more in U.S. human capital than natives, and low-skill-transferability immigrants will invest more than high-skill-transferability immigrants. Low-skill-transferability immigrants will experience higher earnings growth than comparably educated natives. Among immigrants with comparable initial levels of schooling, there will be an inverse relationship between entry earnings and earnings growth.

The IHCI model implies that an immigrant's level of source-country human capital cannot be measured by his or her entry earnings. Entry earnings for immigrants with the same level of source-country human capital will differ both by the degree to which their skills can realize an immediate return in the U.S. labor market (transferability) and by the

and starting again. Testing these alternative perspectives requires examining what happens over time to the earnings and human capital investment of the mismatched immigrants in their new country.

[3] The initial references for the four concepts are as follows: for the first concept: Chiswick (1978, 1979); for the second concept: Duleep and Regets (1994, 1999, 2002); for the third concept (Duleep and Regets 1992); and for the fourth concept, permanence is incorporated in the version presented in Duleep and Regets (1999). With the parameter τ_P, Duleep and Regets (2002) incorporate the proportion of source-country human capital that transfers to the production of new, destination-country human capital.

degree to which entry earnings are diminished by U.S. human capital investment. As such, the adjusted entry earnings of immigrants is a misguided measure of immigrant quality.

With its emphasis on the low opportunity cost of human capital investment for immigrants lacking transferable skills paired with the value of source-country human capital for learning new skills, a distinguishing feature of the IHCI model is that the higher incentive to invest in human capital pertains not only to host-country-specific human capital that restores the value of specific source-country human capital (the foreign-born aeronautical engineer who learns English so that he can pursue aeronautical engineering again) but to new human capital investment in general.

U.S. natives well launched in their careers or immigrants with highly transferable skills that allow them to immediately pursue jobs in their fields would be reluctant to pursue training outside of their fields. The low opportunity cost for similarly educated immigrants who could not initially transfer their source-country human capital, paired with the value of that human capital for learning new skills, may make pursuing new directions an attractive option. The higher propensity to invest in human capital, beyond that which restores the value of immigrants' original human capital, makes low-skill-transferability immigrants a flexible and dynamic component of their new country's labor force.

Measuring Skill Transferability

The IHCI model assumes that the adjusted earnings gap—the difference between an immigrant's earnings and that of a native of the same level of human capital (e.g., same age and education level)—measures immigrant skill transferability. The larger the gap, the greater the human capital investment, and the higher the earnings growth.

Immigration scholars often use English language proficiency as a measure of skill transferability. Chiswick's (1979) path-breaking work on skill transferability and earnings growth compared groups coming from English-speaking versus non-English-speaking countries. Seeking a more direct measure of skill transferability than the adjusted earnings gap, Akresh (2007) and Borjas (2015) also use English proficiency.

A problem with using English proficiency to measure skill transferability is that the relationship between English proficiency and skill transferability does not occur in a vacuum. Non-English-speaking immigrants may seek U.S. employment that does not require English fluency. English-speaking immigrants may have low skill transferability despite being fluent in English if (for instance) their migration necessitates embarking on a new line of work. These thoughts suggest that English proficiency will not consistently measure immigrant skill transferability. Moreover, English proficiency captures but one aspect of skill transferability whereas the adjusted earnings gap captures all aspects of skill transferability, measured and unmeasured.

Concluding Remark

With their lower adjusted entry earnings, hence likely lower skills transferability, we would expect that Asian immigrants from countries that are less economically developed than the United States would have higher earnings growth than immigrants from Western Europe, Canada, and Japan.

References

Akresh, I.R. 2007. "U.S. Immigrants' Labor Market Adjustment: Additional Human Capital Investment and Earnings Growth." *Demography*, 44(4): 865–881.

Borjas, 2015. "The Slowdown in the Economic Assimilation of Immigrants: Aging and Cohort Effects Revisited Again," *Journal of Human Capital*, 9(4), 483–517.

Chiswick, B.R. (1978) "The Effect of Americanization on the Earnings of Foreign-Born Men," *Journal of Political Economy*, October, pp. 897–922.

Chiswick, B.R. (1979) "The Economic Progress of Immigrants: Some Apparently Universal Patterns," in W. Fellner, ed., *Contemporary Economic Problems*. Washington, D.C.: American Enterprise Institute, pp. 359–99.

Chiswick, B.R. (1980) *An Analysis of the Economic Progress and Impact of Immigrants*, Department of Labor monograph, N.T.I.S. No. PB80-200454. Washington, DC: National Technical Information Service.

Duleep, H. and Regets, M., (1992) "Some Evidence on the Effect of Admission Criteria on Immigrant Assimilation: The Earnings Profiles of Asian Immigrants in Canada and the U.S" in *Immigration, Language and Ethnic Issues: Canada and the United States*, Barry Chiswick (ed.). Washington, DC: American Enterprise Institute, 1992, pp. 410–437.

Duleep, H. and Regets, M. (1994) "The Elusive Concept of Immigrant Quality: Evidence from 1960–1980," (1992 American Economic Association version), Working Paper PRIP-UI-28, Washington, DC: Urban Institute.

Duleep, H. and Regets, M., (1999) "Immigrants and Human Capital Investment," *American Economic Review*, May, pp. 186–191.

Duleep, H. and Regets, M.,. (2002) "The Elusive Concept of Immigrant Quality: Evidence from 1970–1990." IZA Discussion Paper No. 631. Institute for the Study of Labor, Bonn, Germany.

Mayer, R. E., & Wittrock, M. C. (1996). "Problem-solving transfer." In D. C. Berliner & R. C. Calfee (Eds.), *Handbook of educational psychology* (p. 47–62). Macmillan Library Reference Usa; Prentice Hall International.

Piracha, M., Tani, M. and Vadean, F. (2012) "Immigrant Over- and Under-education: The Role of Home Country Labour Market Experience," *IZA Journal of Migration*, 1.

CHAPTER 5

Methodological Implications of a Human Capital Investment Perspective

A human-capital-investment perspective prescribes an empirical strategy for measuring immigrant economic assimilation. In particular, it suggests that analysts allow earnings growth to vary with entry earnings, examine each year-of-entry cohort separately, limit their analysis to year-of-entry cohorts that can be followed from immigrants' initial years in the host country, and avoid excluding students, the self-employed, or zero earners in studies that follow cohorts instead of individuals.

Constraints on the Relationship Between Entry Earnings and Earnings Growth

Whether with pooled cross-sections (where analysts follow year-of-entry immigrant cohorts across cross-sections from multiple years) or longitudinal data on individuals, economists typically pool immigrants who have entered the host country at different points in time and estimate a variant of the following model:

$$\log y_i = X'\beta + \gamma' C_j + \alpha' \text{YSM} + \varepsilon_i$$

where y_i denotes the earnings of immigrant i; X is a vector of variables measuring education and experience, and β the corresponding coefficients; YSM measures years since migration; and C_j is a set of dummy variables representing each year-of-immigration category, j. This is the fixed-cohort-effect method for measuring immigrant earnings trajectories, which Chap. 1 compared with the cross-sectional method. The coefficients on C_j measure how high entry earnings are for each cohort j while the coefficient on YSM measures earnings growth.

Controlling for levels of human capital that individuals possess, the fixed-cohort-effect method assumes that changes in entry earnings are not accompanied by changes in the earnings growth of immigrant cohorts. In contrast, a human-capital-investment perspective predicts that decreases in entry earnings are *systematically* accompanied by increases in earnings growth, and vice versa.

In practice, the fixed-cohort-effect strategy averages the earnings growth rates of all of the year-of-entry cohorts. With an inverse relationship between entry earnings and earnings growth, the actual earnings growth of a recent cohort with lower (adjusted) initial earnings than earlier cohorts will be higher than the preceding cohorts' earnings growth; an estimate based on the average of the earnings' growth rates of all the year-of-entry cohorts underestimates its earnings growth. The actual earnings growth of a recent cohort with higher initial earnings than previous cohorts will be below the preceding cohorts' earnings growth; the average of the earnings' growth rates overestimates its earnings growth.

This qualification holds whether the analyst's focus is the study of all immigrants—where changes in initial earnings may arise from changes in the source-country composition of immigration—or the study of immigrants from a single country—where inter-cohort changes in entry earnings may reflect changes in that source country's level of economic development vis-à-vis the United States or the admission program its emigrants use to become U.S. immigrants. If cohorts that vary in their entry-level earnings systematically vary in their earnings growth, each year-of-entry cohort should, at least initially, be separately examined. Doing so frees the estimation of each earnings trajectory from the trajectories of other cohorts.

Economists also typically incorporate all possible year-of-entry cohorts in their analysis, including those for which the available earnings information begins years after an immigrant's initial year of entry. Yet, to avoid implicit assumptions about the relationship between entry earnings and earnings growth, analysts should only include year-of-entry cohorts that can be followed from the immigrant's initial years in the host country. While reducing the information that is used, this approach ensures that conclusions are not the result of an assumed relationship between immigrant's entry earnings and earnings growth.[1]

A human-capital-investment perspective further suggests interpreting with caution estimated returns to schooling and experience. Since low-skill-transferability immigrants will be more engaged in human capital investment than high-skill-transferability immigrants, the estimated returns to *cumulative* levels of schooling and experience in an earnings regression will (at least in the initial years of immigration) be much lower for low-skill-transferability immigrants even though the return for schooling that low-skill-transferability immigrants undertake in the host country should, theoretically, exceed that of their high-skill-transferability statistical twins. The return to existing schooling from their home country will be underestimated because it will become more valuable with time in the United States. And the return to U.S. education will likely be even greater than it is for natives because of the way it makes home country skills more usable.

In fact, the very practice of controlling for current education level likely depresses estimates of immigrant earnings growth. Those who pursue more education have their post-education earnings compared not to the earnings of a similar person holding their old education level, but to someone who already held the new education level in the previous period. The earnings growth estimate of someone who did not pursue further education may be biased downwards as well since their later earnings will be grouped with those who had just recently obtained the same education level.

This downward bias will be greatest for immigrants who have low initial earnings given their age and education level. Due to initial problems transferring human capital, they will invest in more schooling than either other immigrants or natives with the same age and education.[2] Their earnings

---

[1] This approach also avoids confounding effects of age and assimilation and the choice of an appropriate reference group (see, for instance Kossoudji 1989; Lalonde and Topel 1991).

[2] In his 1986 paper, Chiswick (p. 188) notes the bias created in earning growth estimates when controlling for education and comparing two groups with different rates of school attendance:

> ... the earnings analyses here and in Borjas (1985) bias downward the cohort increase in earnings over the decade by controlling for schooling level in the same year as the earnings data, rather than schooling level in 1970. While this downward bias occurs for all groups, it is likely to be more intense for Cuban and other refugees as they

growth will be underestimated in both absolute terms and relative to other immigrants or natives.

Because of the downward bias that using current education levels causes, analysts who want to control for education in their earnings estimations should ideally use education levels measured close to an immigrant's year of entry. It is possible to do this in analyses that follow the same individuals. With data following cohorts across cross-sectional datasets, the bias problem may only be ameliorated by using broad categories to measure the education levels of individuals.

Constraints on the Sample

A human-capital-investment perspective cautions against the common practice by economists of excluding from the sample individuals with zero earnings, students, and the self-employed.[3] Excluding zero earners hides dimensions of immigrant assimilation involving human capital investment: it omits immigrants who are pursuing job search and learning, instead of earning. Excluding such individuals likely understates immigrant earnings growth in general and in particular for groups with low initial skill transferability, while having less effect on earnings growth estimates for immigrant groups characterized by high initial skill transferability.

Excluding the self-employed may also understate immigrant earnings growth, particularly for groups with low initial skill transferability. Gallo and Bailey (1996), Bailey (1987), Portes and Bach (1985), Waldinger (1989), and Chunyu (2011) document an immigrant sector in various industries characterized by mutually beneficial arrangements in which recent immigrants working as unskilled laborers at low wages (or even no wages) in immigrant-run businesses receive training and other forms of

---

invest in more post immigration schooling. Thus the downward bias in the estimated growth of earnings would be greater for the Cubans than for other whites.

---

[3] This is, of course, standard professional practice for labor economists estimating the rate of return to education and experience from Mincer earnings functions. Excluding the self-employed, for example, has a practical econometric tradeoff; to the extent that the self-employed have different unmeasured characteristics, excluding them introduces a sample selection bias, but also removes returns to physical and financial capital from reported earnings.

support that eventually lead to more skilled positions or self-employment.[4]

Analyses that use as the dependent variable hourly wages or weekly earnings, instead of annual earnings, omit an important component of economic assimilation—the ability to work longer hours and weeks.

Biases Created by Sample Constraints

In addition to hiding dimensions of immigrant assimilation involving human capital investment, sample constraints can bias estimates of earnings growth in studies that follow immigrant cohorts across censuses or other cross-sectional datasets.

There is an obvious bias created in estimates of cohort earnings growth if you exclude from the first period individuals with zero earnings, but include individuals who previously had zero earnings in the second period. Even if the typical ten years between measurements was enough for those with initial skill transferability problems to earn the same as immigrants with no initial difficulty, their exclusion from the first period downward biases earnings growth estimates. This bias also distorts inter-cohort comparisons of how earnings growth changes as adjusted entry earnings change. The greater the difficulty immigrants have transferring their human capital to the U.S. labor market, the greater the incentive for them to invest in human capital. Yet, the greater the tendency for immigrants to invest in human capital, the greater the presence of zero earners in the first period, and the stronger the potential for a negative bias caused by excluding zero earners from the cross-sections over which earnings growth is being measured.

More generally, for studies that follow immigrant cohorts across two or more cross-sections, sample-selection rules such as excluding zero earners, students, and the self-employed compromise the comparability of the two census samples.[5] Immigrants excluded from the initial census sample because they are unemployed or out of the labor force (perhaps because of job search or time spent in school) may be fully employed, hence included,

---

[4] Lofstrom (2002) finds that including the self-employed reduces the immigrant-native earnings gap by 14%.

[5] Sample comparability is not an issue when analysts follow the same individuals with longitudinal data.

in the second census sample; immigrants also move from wage and salary jobs (thereby included in the first census sample) to self-employment (thereby excluded). Sample comparability issues apply to any cohort followed between censuses (or other data sources), but are particularly important for the study of immigrant economic assimilation since immigrants have high occupational mobility, high in-school rates, and a high propensity to become self-employed.[6]

Beyond the effect of sample constraints on immigrant assimilation per se, most studies measure the earnings growth of immigrants *relative* to U.S. natives. In studies that follow cohorts instead of individuals, sample constraints distort the estimated relative earnings growth of immigrants because the constraints differentially affect immigrants and natives (Duleep et al. 2020).[7]

Because sample restrictions hide dimensions of immigrant economic assimilation and create bias for studies that follow cohorts across cross-sections, we advocate including the self-employed, imposing no labor-force-status restrictions, and meeting the concerns that prompted these sample restrictions in other ways. For instance, rather than excluding the self-employed, analysts could reduce the effect of non-labor income flows on measured earnings by measuring earnings at the median instead of the mean. Excluding the self-employed comes from a desire to measure earnings from work per se versus returns from capital. Yet, if the main policy interest is to measure the economic contribution of immigrants, it matters little if a portion of their increased income comes from their investment of savings into a business whereas excluding those creating businesses may distort the overall picture of immigrant economic assimilation.

Concluding Remarks

Since factors that affect entry earnings also affect earnings growth, the earnings trajectories of year-of-entry cohorts—even from the same source country—may require separate analysis, or at least modeled to permit

---

[6] For occupational mobility, see Akresh (2006, 2008), Chiswick (1978), Chiswick, Lee, and Miller (2005), Chiswick and Miller (2008, 2009), Green (1999), Jasso and Rosenzweig (1990, 1995), and Zorlu (2013). For educational investment see, for instance, Duleep and Regets (1999), Chiswick and DebBurman (2004), Van Tubergen and van de Werfhorst (2007), and Jasso and Rosenzweig (1990).

[7] Sample constraints such as excluding zero earners *can* be imposed if following the same individuals. See, for instance, Duleep and Regets (1997) and Duleep and Dowhan (2002).

factors that affect entry earnings to also affect earnings growth. In the ensuing analyses that follow cohorts across censuses, we analyze each cohort separately and include only year-of-entry cohorts that can be followed from the cohort's initial years in the United States. In addition to our free-form methodology for measuring earnings growth, our analyses that follow cohorts include the self-employed, students, and individuals with zero earnings.

If low-skill-transferability immigrants are more likely to invest in schooling than natives or high-skill-transferability immigrants, then earnings analyses that control for years of schooling, other than pre-immigration schooling, will understate the earnings growth of immigrants who start their U.S. lives with relatively low earnings. As there is no way to control for pre-immigration education in following cohorts across censuses, when we do control for years of schooling in our earnings analyses, we do so within broad categories.

References

Akresh, I.R. 2006. "Occupational Mobility among Legal Immigrants to the United States." *International Migration Review*, 40(4).

Akresh, I.R. 2008. "Occupational Trajectories of Legal U.S. Immigrants: Downgrading and Recovery." *Population and Development Review*, 34(3): 435–456.

Bailey, Thomas R. (1987) *Immigrant and Native Workers: Contrasts and Competition*, Conservation of Human Resources, Boulder and London: Westview Press.

Borjas GJ. 1985. "Assimilation, Changes in Cohort Quality, and the Earnings of Immigrants." *Journal of Labor Economics* 3: 463–489.

Chiswick, 1986. "Is the New Immigration Less Skilled than the Old?" *Journal of Labor Economics*, 4(2):168–192

Chiswick, B.R. (1978) "A Longitudinal Analysis of Occupational Mobility of Immigrants," in ed. Barbara Dennis, *Proceedings of the 30th Annual Winter Meeting, Industrial Relations Research Association*, December, 1977, Madison, Wisconsin, 1978b, pp. 20–27.

Chiswick, Barry and Noyna DebBurman. 2004. "Educational Attainment: Analysis by Immigrant Generation," *Economics of Education Review* 23:361–379.

Chiswick, B.R., Lee, Y.L., and Miller, P.W. (2005) "Longitudinal Analysis of Immigrant Occupational Mobility: A Test of the Immigrant Assimilation Hypothesis." *International Migration Review*, 39(2):332–353.

Chiswick, B.R and Miller, P.W. (2008) "Occupational Attainment and Immigrant Economic Progress in Australia," *Economic Record*, 2008, 84, S45-S56

Chiswick, B.R. and Miller, P.W. (2009) "Earnings and Occupational Attainment among Immigrants" *Industrial Relations*, 48 (3), 454–465.

Chunyu, Miao David. 2011. "Earnings Growth Patterns of Chinese Labor Immigrants in the United States,"Paper presented at the 2011 Annual Meeting of the Population Association of America, Washington, D.C. http://paa2011.princeton.edu/download.aspx?submissionId=110457

Duleep, H. and Dowhan, D. (2002)."Insights from Longitudinal Data on the Earnings Growth of U.S. Foreign-born Men," *Demography*, August.

Duleep, H. and Regets, M. (1997) "Measuring Immigrant Wage Growth Using Matched CPS Files," *Demography*, vol. 34, May, pp. 239–49.

Duleep, H. and Regets, M., (1999) "Immigrants and Human Capital Investment," *American Economic Review*, May, pp. 186–191.

Duleep, Harriet & Liu, Xingfei & Regets, Mark, 2020. "How the Earnings Growth of U.S. Immigrants Was Underestimated," Working paper.

Gallo C, Bailey TR. (1996) "Social Networks and Skills-Based Immigration Policy," in *Immigrants and Immigration Policy: Individual Skills, Family Ties, and Group Identities*, HO Duleep, PV Wunnava, eds. Greenwich, Conn.: JAI Press.

Green DA. 1999. "Immigrant Occupational Attainment: Assimilation and Mobility Over Time." *Journal of Labor Economics* 17: 49–79.

Jasso G, Rosenzweig MR. 1990. *The New Chosen People: Immigrants in the United States*. New York: Russell Sage Foundation.

Jasso G, Rosenzweig M. R. 1995. "Do Immigrants Screened for Skills Do Better than Family-Reunification Immigrants?" *International Migration Review* 29: 85–111.

Kossoudji, Sherrie A., "Immigrant Worker Assimilation: Is It a Labor Market Phenomenon?" *Journal of Human Resources*, Vol 24, No.3, Summer 1989, pp 494–527.

LaLonde, R. J. and Topel, R. H. (1991) "Immigrants in the American Labor Market: Quality, Assimilation, and Distributional Effects," *The American Economic Review*, May.

Lofstrom, Magnus, (2002) "Labor Market Assimilation and the Self-Employment Decision of Immigrant Entrepreneur," *Journal of Population Economics*, 15 (1), 83–114

Portes A and Bach R. (1985) *Latin Journey: Cuban and Mexican Immigrants in the United States*. Berkeley, Calif.: University of California Press.

Van Tubergen F and Van de Werfhorst H. (2007) "Post-immigration Investments in Education: A Study of Immigrants in the Netherlands," *Demography*. Nov: 44(4):883–98.

Waldinger R. (1989) "Structural Opportunity or Ethnic Advantage? Immigrant Business Development in New York." *International Migration Review* 23: 48–72.

Zorlu, A. (2013). "Occupational Adjustment of Immigrants," *Journal of International Migration and Integration*, 14 (4), 711–731.

PART II

Earnings Growth and Human Capital Investment of Immigrant Men, the 1965–1970 and 1975–1980 Cohorts

Asian immigrant men start their U.S. lives at much lower earnings than their comparably educated counterparts from Europe and Canada (Table 1.2). One might naturally assume that once analysts have adjusted for differences in education and age, the earnings growth of an earlier cohort will predict the earnings growth of a more recent cohort. This

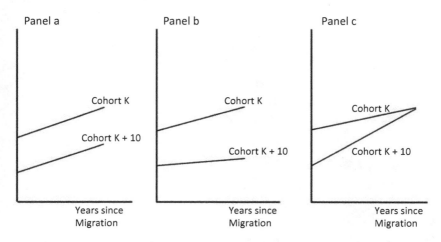

Fig. II.1 Earnings growth scenarios. (Source: Authors' creation)

assumption, pictured in panel A of Fig. II.1, is embedded in the fixed-cohort effect methodology. It implies that immigrant cohorts with low entry earnings will forever remain at a substantial earnings disadvantage relative to U.S. natives, and relative to immigrant cohorts with high initial earnings. The situation worsens if entry earnings and earnings growth are positively related (panel B). This would occur if unmeasured productivity characteristics dampen both initial earnings and earnings growth. If, however, variations in skills transferability underlie variations in the education- and age-adjusted entry earnings of immigrants, then we would expect an inverse relationship between immigrants' entry earnings and their subsequent earnings growth (panel C).

As we compare the earnings profiles of Asian and European/Canadian immigrants, and observe how these earnings profiles change over time, we will be exploring which panel in Fig. II.1 most accurately depicts the relationship between immigrant entry earnings and earnings growth.

CHAPTER 6

The Earnings Growth of Asian Versus European Immigrants

Earnings growth signifies human capital investment. If low-skill-transferability immigrants invest more in U.S. human capital than comparably educated and experienced high-skill-transferability immigrants (or U.S. natives) then, once we have controlled for immigrant's initial levels of schooling and experience, there should be an inverse relationship between immigrant entry earnings and earnings growth. To test this prediction, we use Public Use Microdata Samples (PUMS) of the 1970, 1980, and 1990 censuses to measure the entry earnings and earnings growth over ten years of two cohorts of Asian and European immigrants—those who immigrated at the start of the 1965 immigration reform, and those who immigrated ten years later.[1]

Following the guidelines of Chap. 5, we pursue a simple non-parametric approach that avoids assumptions about the functional form of the earnings profile. Specifically, we use 1970 census data to measure the entry earnings of immigrant men who report coming to the United States between 1965 and 1970 (the 1965–1970 cohort) and we use 1970 and 1980 census data to measure their earnings growth over ten years. We use 1980 census data to measure the entry earnings of the 1975–1980 cohort and 1980 and 1990 data to measure their ten-year earnings growth.

---

[1] This work builds on a general inverse relationship established across all source countries and over time using 1970–1990 census data (Duleep and Regets 1996, 1997a, b) and using 1960–1980 census data (Duleep and Regets 1992, 1994a, b, 1997b).

© The Author(s), under exclusive license to Springer Nature Switzerland AG 2020
H. Duleep et al., *Human Capital Investment*,
https://doi.org/10.1007/978-3-030-47083-8_6

We include zero earners, students, and the self-employed in our samples.[2] Sample size concerns, and our concern about underestimating earnings' growth by controlling for education (discussed in Chap. 5), led us to broadly categorize age and education. The education categories are 1–12 years and 13 or more years; the age categories are 25–39 and 40–54 in the initial census sample and 35–49 and 50–64 in the subsequent census sample.

The earnings at entry and ten years later are measured at the median within education and age subsets. The median is a much less volatile measure of central tendency than the mean in small samples. Using the median also circumvents the problem that census earnings data are truncated. The earnings growth of each region/age/education group for the 1965–1970 cohort is the difference between the 1970 earnings and earnings measured ten years later by the 1980 census, divided by the 1970 earnings; for the 1975–1980 cohort, it is the difference between the 1980 earnings and earnings measured ten years later by the 1990 census, divided by the 1980 earnings.[3]

THE INVERSE RELATIONSHIP: ASIAN AND EUROPEAN IMMIGRANTS

With one exception, in each age/education category, for both the 1965–1970 and 1975–1980 cohorts, Asian immigrant men have lower entry earnings but higher earnings growth than their European counterparts (Table 6.1). This concurs with the IHCI Model if, as seems likely, those who migrate from Europe generally have fewer problems transferring their human capital to the United States than do Asian immigrants

---

[2] Self-employment earnings do usually include some return on financial or physical capital that would bias regression coefficients of rates of return. Including these income flows in an assessment of immigrant economic contributions is not problematic. Moreover, our use of the median rather than the mean of earnings reduces the effect of non-labor income flows from the self-employed; we interpret changes in earning over time as primarily resulting from human capital investment.

[3] An alternative approach to the one we pursue would be to first estimate a parametric model and then, using the predicted values from this model, estimate the correlation between the predicted entry earnings and predicted earnings growth. Although our approach does not make use of information beyond the median within each age/education/country-of-origin cell, the advantage of the approach we pursue is that we can be very certain that our results are not due to a particular set of model assumptions.

Table 6.1 Entry earnings and ten-year real earnings growth rates of age-education cohorts by region of origin (Median earnings, 1989 dollars, deflated by index of weekly wages)

Year of Entry	Immigrants from Asia				Immigrants from Europe			
	1965–1970		1975–1980		1965–1970		1975–1980	
	Entry earnings	Earnings growth	Entry earnings	Earnings growth	Entry earnings	Earnings growth	Entry earnings	Earnings growth
25–39 years old; 1–12 years of school	$12,868	34.1	9887	92.2	21,129	11.4	15,690	58.5
25–39 years old; more than 12 years of school	16,045	134.6	12,553	154.9	27,167	44.3	23,531	70.0
40–54 years old; 1–12 years of school	11,756	20.1	9417	32.3	19,223	-2.1	15,690	21.1
40–54 years old; more than 12 years of school	30,662	-2.8	17,258	42.7	31,933	8.6	29,019	12.0

Source: Authors' estimates based on a 1970 Census of Population 6% microdata sample created by combining six 1% Public Use Sample files, a 6% 1980 file created by combining and reweighting the 5% and 1% files of the 1980 Census of Population, and a 6% microdata 1990 sample created by combining and reweighting the 1990 Census of Population Public Use 5% and 1% Public Use samples. Appendix A provides information on sample sizes for entry cohorts at entry and ten years later

Notes: For the three 1970 1% Census Files with no year-of-immigration data, residing abroad five years before is used as an indicator of being in the 1965–1970 cohort. Foreign born are defined as persons born outside of the United States excluding those with U.S. parents. The sample does not exclude students, the self-employed, and persons with zero earnings

who (with the exception of the Japanese) come from countries that are less economically developed than the United States. For both Asian and European immigrants, as entry earnings from the earlier to more recent cohort decrease, earnings growth increases.

Differences between high- and low-education cohorts in their earnings growth rates further buttress the IHCI Model. The difference is large for younger cohorts, but much smaller and sometimes negative for older entry cohorts. Following an earlier intuition in Chap. 4, source -country education or other human capital increases both the costs and returns to further human capital investment. The additional time the young have to realize the return on their investment increases the lifetime earnings effect of the interaction of prior human capital and new human capital production, while the effect of prior human capital on the cost of investment is unaffected by age. European immigrants have smaller differentials between the earnings growth rates of high- and low-educated immigrants than do Asian immigrants. This concurs with the IHCI model's prediction that the lower the level of skill transferability, the more likely it is that higher education leads to greater human capital investment.

Table 6.2 follows the earnings of Asian and European immigrants relative to U.S. natives, and separately shows the earnings of immigrants from Western Europe. Whether from Europe or Asia, immigrant men of both the 1965–1970 and 1975–1980 cohorts have earnings after 10 to 15 years in the United States that equal or surpass U.S. natives' earnings. This is true despite Asian immigrants' much lower starting point: for the 1965–1970 cohort, Asians initially earned 59% of U.S. natives' earnings versus 83% for the Europeans; for the 1975–1980 cohort, Asians initially earned 50% versus 77% for the Europeans. For both Asian and European immigrants, relative entry earnings fell between the 1965–1970 and 1975–1980 entry cohorts. Yet, earnings growth increased.

The Fixed-Cohort-Effect Model Revisited

According to the fixed-cohort-effect model, once we have controlled for the education and experience level of immigrants, analysts can capture unmeasured factors that affect earnings and vary across entry cohorts by including in the estimation a dummy variable for each entry cohort. This allows the height of the estimated earnings trajectory to vary but implicitly assumes that earnings growth remains constant over year-of-entry cohorts.

Table 6.2 Entry earnings and earnings after ten years, relative to native born of age-education cohorts by region of origin, men 25–54 years old (Median earnings)

	1965–70 cohort		1975–80 cohort		
	Entry earnings, 1970 census	Earnings 10 years later, 1980 census	Entry earnings, 1980 census	Earnings 10 years later, 1990 census	Predicted earnings with fixed-cohort effect assumption
Immigrants from Asia	0.59	1.25	0.50	1.01	
25–39 years old; 1–12 years of school	0.51	0.65	0.45	0.95	0.57
More than 12 years of school	0.46	0.94	0.42	0.94	0.86
40–54 years old; 1–12 years of school	0.44	0.69	0.36	0.83	0.56
More than 12 years of school	0.60	0.76	0.41	0.76	0.52
Immigrants from Europe	0.83	1.00	0.77	1.15	
25–39 years old; 1–12 years of school	0.83	0.88	0.71	1.24	0.74
More than 12 years of school	0.86	0.97	0.79	1.16	0.90
40–54 years old; 1–12 years of school	0.76	0.92	0.60	1.27	0.73
More than 12 years of school	0.77	0.88	0.69	0.98	0.78
Immigrants from Western Europe	0.83	0.99	0.90	1.21	

(*continued*)

Table 6.2 (continued)

	1965–70 cohort		1975–80 cohort		
	Entry earnings, 1970 census	Earnings 10 years later, 1980 census	Entry earnings, 1980 census	Earnings 10 years later, 1990 census	Predicted earnings with fixed-cohort effect assumption
25–39 years old; 1–12 years of school	0.82	0.85	0.79	1.25	0.82
More than 12 years of school	0.82	1.01	0.89	1.24	1.10
40–54 years old; 1–12 years of school	0.71	0.88	0.71	1.20	0.88
More than 12 years of school	0.69	0.97	1.00	1.01	1.41

Source: Authors' estimates based on a 1970 Census of Population 6% microdata sample created by combining six 1% Public Use Sample files, a 6% 1980 file created by combining and reweighting the 5% and 1% files of the 1980 Census of Population, and a 6% microdata 1990 sample created by combining and reweighting the 1990 Census of Population Public Use 5% and 1% Public Use samples. Appendix A provides information on sample sizes for entry cohorts at entry and ten years later

Notes: For the three 1970 1% Census Files with no year-of-immigration data, residing abroad 5 years before is used as an indicator of being in the 1965–1970 cohort. Native born are defined as persons born in the U.S.; foreign born are defined as persons born outside of the United States excluding those with U.S. parents. The sample does not exclude students, the self-employed, and persons with zero earnings

The last column of Table 6.2 illustrates the consequences of assuming stationary earnings growth when the adjusted entry earnings of immigrants are changing: we predict earnings after ten years assuming that the earnings growth of the 1975–1980 cohort is the same as the earnings growth of the 1965–1970 cohort from the same source region. Comparing the last column with the next-to-last column shows that when entry earnings decrease between the first and second cohort, the fixed-cohort-effect assumption underestimates the earnings growth, hence the earnings after ten years, of immigrant men for each age and education group. This is because decreases in entry earnings (for both Europeans and Asians) accompany increases in earnings growth.

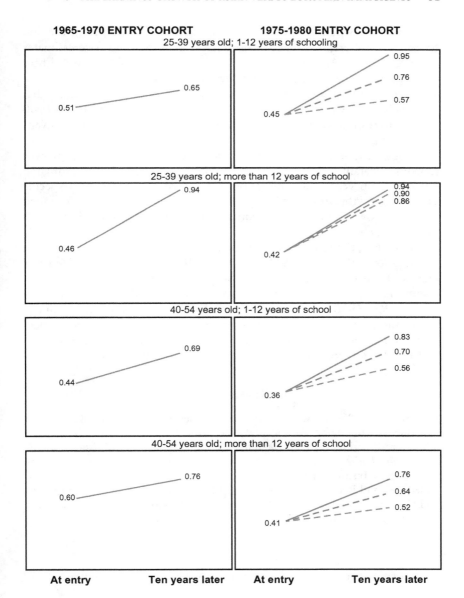

Fig. 6.1 Actual and predicted earnings trajectories for Asian immigrants for 1965–1970 entry cohort by age and years of schooling. Foreign-born/U.S.-born median earnings at entry and ten years later. (Source: Authors' creation)

Conversely, when the (adjusted) entry earnings increase from the first to second cohort, the fixed-cohort-effect methodology overestimates earnings growth. An example of this is in the last row of Table 6.2. Entry earnings increased substantially for West European immigrants who were at least 40 years old with more than 12 years of schooling.[4] Assuming that the earnings growth of the 1965–1970 cohort applies to the 1975–1980 cohort overestimates the most recent cohort's earnings growth and eventual earnings.

Though the fixed-cohort-effect model assumes that earnings growth remains constant across year-of-entry cohorts, empirically the estimated coefficient on years since migration is an average of the cohort-specific earnings growth rates. For any given cohort, how closely this averaged earnings growth estimate approximates actual earnings growth depends on how much entry earnings vary across the year-of-entry cohorts.

In Fig. 6.1, the blue line shows the actual earnings trajectories of Asian immigrants in the 1965–1970 and 1975–1980 entry cohorts. The silver line is the earnings trajectory predicted using the fixed-cohort-effect assumption of stationary earnings growth: we apply the earnings growth of the 1965–1970 cohort to the 1975–1980 cohort. The orange line is the earnings trajectory from an estimation using the fixed-cohort-effect methodology in which the average earnings growth of the 1965–1970 and 1975–1980 cohorts is applied to the 1975–1980 cohort.

References

Duleep, H. and Regets, M., (1992) "Some Evidence on the Effect of Admission Criteria on Immigrant Assimilation: The Earnings Profiles of Asian Immigrants in Canada and the U.S" in *Immigration, Language and Ethnic Issues: Canada and the United States*, Barry Chiswick (ed.). Washington, DC: American Enterprise Institute, 1992, pp. 410–437.

Duleep, H. and Regets, M. (1994a) "The Elusive Concept of Immigrant Quality: Evidence from 1960–1980," (1992 American Economic Association version), Working Paper PRIP-UI-28, Washington, DC: Urban Institute.

---

[4] We define Western Europe as Europe minus the former Soviet Union countries.

Duleep, H. and Regets, M. (1994b) "Country of Origin and Immigrant Earnings," Working Paper PRIP-UI-31, Washington, DC: Urban Institute.

Duleep, H. and Regets, M. (1996) "Earnings Convergence: Does it Matter Where Immigrants Come From or Why?," *Canadian Journal of Economics*, vol. 29, April.

Duleep, H. and Regets, M. (1997a) "The Decline in Immigrant Entry Earnings: Less Transferable Skills or Lower Ability?" *Quarterly Review of Economics and Finance*, vol. 37, Special Issue on Immigration, pp. 189–208.

Duleep, H. and Regets, M. (1997b). "Immigrant entry earnings and human capital growth." *Research in Labor Economics*, 16, 297–317.

CHAPTER 7

The Earnings Profiles of Immigrant Men in Specific Asian Groups: Cross-Sectional Versus Cohort-Based Estimates

The comparisons thus far have included *all* Asians, including refugees. Moreover, immigrants from Japan, an economically developed country are combined with Asian immigrants from economically developing countries. This chapter examines the economic assimilation of immigrant men in the non-refugee groups that dominated post-1965 U.S. Asian immigration.[1] Ranked by their 1980s immigrant population size, they are the Filipinos, Chinese (including entrants from China, Taiwan, and Hong Kong), Koreans, Indians, and Japanese.[2]

Initial Earnings

Using microdata samples from the 1970 and 1980 censuses, Table 7.1 presents the median earnings of Asian immigrant men, 25–64 years old, who have been in the United States for five or fewer years, alongside the earnings of European and Canadian recent arrivals. Western Europe (further broken down into English-speaking—the United Kingdom and Ireland—and non-English-speaking) includes all European countries other than the former Soviet Union countries. The ratios of each group's median entry earnings to the median earnings of working-age U.S.-born men are shown in parentheses.

---

[1] Please refer to Reimers (2005, pp. 157–206).
[2] Authors' estimates based on 1980 Census PUMS.

The regional variations among our comparison groups conform to expectations based on intergroup variations in skill transferability. Immigrants from Canada and English-speaking Europe have the highest initial earnings; within Western Europe, immigrants from the U.K. and Ireland earn more than do immigrants from non-English-speaking Western Europe. Immigrants from the former Soviet Union countries (also referred to as Eastern Europe) earn far less than do their Western European and Canadian counterparts. This result conforms both to the hypothesis that immigrant skills transfer more easily when source- and host country economic systems are similar and to the hypothesis that

Table 7.1 Median entry earnings of 25–64-year-old male immigrants, 1989 dollars (percentage of U.S. native born in parentheses)

	1965–1970 entry cohort; 1969 earnings	1975–1980 entry cohort; 1979 earnings
Philippines	14,582 (56%)	14,402 (57%)
China	10,416 (40)	9131 (36)
Korea	19,390 (75)	14,199 (56)
India	19,711 (76)	17,223 (68)
Japan	22,595 (87)	27,562 (109)
Western Europe	21,794 (84)	23,321 (92)
English speaking	32,211 (124)	31,261 (123)
Other Western Europe	20,031 (74)	20,280 (77)
Eastern Europe	22,596 (83)	13,523 (52)
Canada	29,327 (108)	30,425 (116)
All Immigrants	17,788 (68)	12,949 (51)

Source: Authors' estimates: For the 1965–1970 cohort, estimates are based on a 1970 Census of Population 6% microdata sample created by combining six 1% Public Use Sample files. For the 1975–1980 cohort, estimates are based on a 6% 1980 file created by combining and reweighting the 5% and 1% files of the 1980 Census of Population. Appendix A provides information on sample sizes by entry cohort

Notes: For the three 1970 1% Census Files with no year-of-immigration data, residing abroad five years before is used as an indicator of being in the 1965–1970 cohort. Native born are defined as persons born in the U.S.; foreign born are defined as persons born outside of the United States excluding those with U.S. parents. China includes mainland China, Taiwan, and Hong Kong. Korea includes South, North, and unspecified Korea. The sample does not exclude students, the self-employed, and persons with zero earnings

differences in economic opportunity and social conditions encourage the immigration of persons lacking U.S.-transferable skills.[3]

Asian recent arrivals from economically developing countries earn far less than do their Western European and Canadian counterparts: the latter earn over 80% U.S. natives' earnings, the former only half. Both absolutely, and relative to their Western European and Canadian counterparts, the entry earnings of the developing country Asian immigrant men fell over time. Educational attainment cannot explain their low entry earnings. Immigrant men in these groups are more likely to have a high school diploma and at least as likely to have a college degree as immigrants from Canada and non-English speaking Western Europe (Table 7.2).[4] Indian immigrants are particularly extraordinary in this regard: over 40% report 18 or more years of schooling.[5]

Nor can other general characteristics explain their low entry earnings. Using coefficients from group-specific regressions estimated on 1980 census data, Table 7.3 shows the earnings of each immigrant group evaluated at U.S.-born averages. In these regressions, the natural logarithm of earnings is regressed on years of schooling (a three-part spline), age, age squared, age × education, years since migration, education × years since migration, marital status, metropolitan status, and U.S. region of residence.[6] The first row shows the entry earnings of each immigrant group evaluated one year after migration and relative to the earnings of

---

[3] See the discussion in Chap. 3 of what causes variation in skill transferability among immigrant groups.

[4] The United Kingdom/Ireland group has an exceptionally low percentage of immigrants with less than a high school education. Among all U.S.-born men (ages 25–64), 26% in 1980 and 18% in 1990 had less than 12 years of schooling. For U.S.-born non-Hispanic white men, 22.4% had less than 12 years of schooling in 1980.

[5] The proportion of college graduates among all of these immigrant groups—Asian, European, and Canadian—surpasses that of U.S. natives of whom 21.8% in 1980 and 25.1% in 1990 reported 16 or more years of schooling. In 1980, 24.5% of U.S.-born non-Hispanic white men, ages 25–64, had 16 or more years of schooling. The educational levels of the four developing Asian source countries that have dominated post-1965 non-refugee Asian immigration to the U.S. far exceed the educational levels of immigrants in general (Table 7.2). In 1980 and 1990, 17.4% and 12.3% of immigrants from the Philippines, China, Korea, and India reported less than 12 years of schooling compared with 36% of all immigrants in those years; 20.5% in 1980 and 26.2% in 1990 of the combined developing Asian group reported 18 or more years of schooling compared with 13.1% and 14.9% in those years for the all-immigrant group.

[6] The estimated coefficients of this model are presented in appendix A of Duleep and Regets (1992). U.S.-born men in this analysis include non-Hispanic white men only.

Table 7.2 Entry education level of the 1975–1979 cohort of 25–54-year-old male immigrants (percent)

	1975–1979 cohort, 1980 years of education					
	Less than 9 years	9–11	12	13–15	16–17	18 or more years
Philippines	11.4	6.7	13.5	23.1	34.5	11.0
China	19.6	7.3	15.8	13.2	24.6	19.5
Korea	5.3	5.8	26.6	17.3	34.9	10.2
India	4.5	6.3	9.6	14.2	21.9	43.5
Japan	2.2	1.5	15.4	12.1	49.6	19.2
Western Europe	20.3	6.7	20.1	15.3	19.1	18.4
English speaking	2.7	6.3	22.2	20.7	26.0	22.1
Other West Europe	28.6	6.9	19.2	12.8	15.9	16.7
Eastern Europe	13.9	13.1	20.9	13.4	23.0	15.8
Canada	3.9	12.5	19.1	20.9	22.3	21.2
All Immigrants	26.7	9.4	18.5	16.0	16.4	13.1

Source: Authors' estimates based on a 6% 1980 file created by combining and reweighting the 5% and 1% files of the 1980 Census of Population. Appendix A provides information on sample sizes by entry cohort

Notes: Foreign born are defined as persons born outside of the United States excluding those with U.S. parents. China includes mainland China, Taiwan, and Hong Kong. Korea includes South, North, and unspecified Korea. The sample does not exclude students, the self-employed, and persons with zero earnings

Table 7.3 Relative earnings of U.S. immigrants, evaluated at U.S. native-born mean level of schooling

	Filipino	Chinese	Korean	Indian	Japanese	European	British
Earnings ratio at 1 year	0.59	0.43	0.54	0.49	1.01	1.07	1.23
Earnings ratio at 10 years	0.89	0.75	0.93	0.92	1.07	0.99	0.99
Average earnings growth during first 10 years	0.07	0.09	0.08	0.09	0.03	0.02	0.006

Source: Authors' estimates based on a 6% 1980 file created by combining and reweighting the 5% and 1% files of the 1980 Census of Population

Table 7.4 English language proficiency of 1975–1980 entry cohort of 25–64-year-old male immigrants (percent)

	1975–1980 entry cohort; 1980 data	
	Speaks English poorly or not at all	Speaks only English or Speaks English very well
Philippines	9.4%	50.9%
China	42.2	19.0
Korea	45.9	15.0
India	6.1	68.0
Japan	26.8	25.3
Western Europe	21.0	59.8
English speaking	0.3	99.2
Other Western Europe	30.7	41.4
Eastern Europe	46.5	22.2
Canada	0.8	94.8
All Immigrants	36.7	37.0

Source: Authors' estimates based on a 6% 1980 file created by combining and reweighting the 5% and 1% files of the 1980 Census of Population. Appendix A provides information on sample sizes by entry cohort

Notes: Foreign born are defined as persons born outside of the United States excluding those with U.S. parents. The sample does not exclude students, the self-employed, and persons with zero earnings. China includes mainland China, Taiwan, and Hong Kong. Korea includes South, North, and unspecified Korea. The sample does not exclude students, the self-employed, and persons with zero earnings

U.S.-born men with comparable characteristics. Whereas Europeans earn at least as much as their U.S.-born statistical twins, the initial earnings of immigrant men from the Asian developing countries are about half the earnings of comparable U.S. natives.

The similarity in the entry earnings of Korean, Indian, Filipino, and Chinese immigrants, adjusting for age and schooling, is surprising given the much higher English proficiency of men from India and the Philippines (Table 7.4). These results are not surprising, however, if intergroup differences in skill transferability stem from immigrant selection based on intercountry differences in social conditions and economic opportunity: the common link among the Asian developing countries are economic and social constraints vis-à-vis the United States.[7]

---

[7] Note that this situation has been changing rapidly in South Korea. The change in their relative level of economic opportunity relative to the U.S. should affect the entry earnings of immigrants to the U.S.

More generally, consider two countries with similar limited economic opportunities (or other constraints) relative to the United States. English is the national language in Country A, whereas few people speak English in Country B. With unconstrained immigration and no earnings uncertainty, immigration from countries A and B will continue until, for the marginal immigrant, the net present value of his earnings in his country of origin equals the net present value of his earnings in the United States, minus migration costs. Persons from the English-speaking country will find it worthwhile to migrate, even if their skills *other than English* are not transferable to the United States. This will be true to a much lesser extent for immigrants from the non-English speaking country. The resulting U.S. immigrant population from Country A will consist of persons who speak English, but who often lack other U.S.-specific human capital. For Country B, the resulting U.S. immigrant population will include a smaller fraction of persons who speak English and have a higher degree of skill transferability on dimensions other than English proficiency.

Of course, many factors other than the expected earnings streams in the home and destination country affect immigration. Nevertheless, this simple example illustrates how countries whose populations have very different levels of U.S.-transferable skills (e.g., English proficiency) could have U.S. immigrant populations with similar entry earnings and earnings growth.

WITHIN-COUNTRY CHANGES IN ENTRY EARNINGS

Table 7.1 underscores that changes in the source-country composition of U.S. immigrants following the 1965 Immigration and Nationality Act contributed to the decline in the entry earnings of U.S. immigrants. In addition, within source country changes in immigrant entry earnings occurred. The initial earnings of immigrants from China, Korea, India, and Eastern Europe decreased whereas those of Western Europe, Canada, and Japan stayed the same or increased.

The 1965 Immigration and Nationality Act eliminated country-specific quotas, which favored West European and Canadian immigration, and introduced a system of restricted and non-restricted admission categories that favored immigrants with family members in the United States. It also authorized employment-based immigration: 20% of the numerically restricted visas were allocated to applicants based on their occupational skills.[8] The very nature of employment-based admissions, where employers hire immigrants with specific skills for specific jobs, ensures that these immigrants have higher skill transferability, hence higher initial U.S. earnings, than family-based immigrants—a prediction confirmed by several studies.[9]

When immigration policy changed in 1965, the national origins legislation had been in effect some 40 years. Potential migrants from countries that had previously faced severe restrictions under the national-origins legislation, such as China, Korea, India, and the Eastern European countries, lacked U.S.-based family members. Unable to immigrate via family ties, qualified migrants gained a U.S. foothold via the employment-based admission categories. Once established, their relatives—with less transferable skills—immigrated via the family-based categories. In this way, the proportion of U.S. immigrants—*from the same country*—lacking skills that immediately transfer to the U.S. labor market, grew in the years following 1965. This scenario may explain why, even holding source country constant, the age- and education-adjusted entry earnings of immigrants from China, Korea, India, and Eastern Europe declined in the post-1965 era.

Earnings Growth

If immigrants from the Philippines, China, Korea, and India initially earn less than their European and Canadian counterparts because they are initially more deficient in human capital specific to the U.S. labor market, then these immigrants should experience higher earnings growth than their European counterparts. If, however, intergroup differences in entry earnings reflect variations in immigrant quality then immigrant men from the Asian developing countries should have relatively low earnings growth.

Using estimated coefficients from the group-specific earnings regressions, we simulate the earnings path of each immigrant group and the

---

[8] The occupational skills (or employment-based) classification embraced two components: workers, skilled and unskilled, in occupations where labor is deemed scarce, and professionals, scientists, and artists of exceptional ability.

[9] Using different data sets, Jasso and Rosenzweig (1995), Duleep and Regets (1996a, b) and DeSilva (1996) find that family-based immigrants start their host-country lives with lower earnings than their employment-based statistical twins but have higher earnings growth; with time, the earnings of the two groups converge.

U.S. born (Table 7.3).[10] Holding years of schooling, marital status, metropolitan status, and region of residence constant at the mean values for U.S.-born men, we compare what immigrant and U.S.-born men would earn if they possessed the same level of education and other general characteristics.[11]

The simulations show that all of the developing-country Asian groups have high earnings growth. After 10 to 15 years in the United States, Filipino, Korean, and Indian immigrant men earn about 90% of the earnings of U.S.-born men with comparable characteristics. In contrast, Japanese, European, and British immigrants—who are characterized by high initial earnings—have low earnings growth. These earnings profiles suggest a story of convergence in line with panel c of Fig. II.1.

COHORT-BASED ESTIMATES OF IMMIGRANT EARNINGS GROWTH

This chapter's exploration of immigrant earnings growth has so far revolved around a cross-sectional analysis. It assumes that for any given group of immigrants, once we control for observed characteristics, such as education and age, each year-of-entry cohort will have a similar earnings trajectory. Yet, we know this assumption is wrong: the adjusted entry earnings of the developing country Asian groups fell over time relative to U.S. natives, and relative to West European and Canadian immigrants. Perhaps Asian immigrants have low earnings growth, and the cross-sectional estimates of high earnings growth stem from a decrease in their entry earnings (Fig. 7.1).

To test the cross-sectional results, we use 1980 and 1990 census data to follow the cohort of immigrant men, ages 25–54, who entered the United States between 1975 and 1980.[12] Not adjusting for any characteristics, the first column of Table 7.5 shows the median earnings of each group during

---

[10] Each simulation begins at age twenty-eight, which for immigrants also serves as the age at migration. For each subsequent year of the simulation, age, age squared, years since migration, years since migration squared, age × education, and years since migration × education are all appropriately incremented and multiplied by their estimated coefficients from the group-specific regressions.

[11] Note that the benchmark of U.S. born men in the analysis presented in Table 7.3 is limited to non-Hispanic white men.

[12] Note that the age and years-since-migration restriction avoids the confounding effects of age and assimilation highlighted in Kossoudji (1989) and Friedberg (1992, 1993).

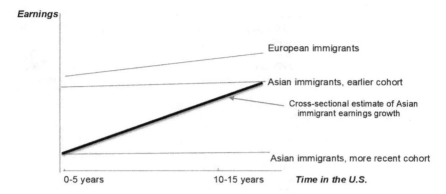

Fig. 7.1 Hypothetical earnings profile of Asian and European immigrants. (Source: Authors' creation)

their first five years in the United States; the second column shows their earnings ten years later. The third column shows the percentage change over the ten-year period. We see, once again, that immigrants from each of the four developing Asian countries have high earnings growth. Immigrants from Eastern Europe (who share the low entry earnings of the developing Asian countries) also exhibit high earnings growth. In contrast, immigrants from Japan, Western Europe, and Canada have relatively low earnings growth.

At the 10- to 15-year point, the earnings of Chinese, Korean, and Filipino immigrant men approach, and the earnings of Asian Indians exceed, the West European mark. The earnings of all of the developing country Asian groups approach or exceed U.S. natives' earnings. Chinese immigrants initially earn only 37% of the U.S. native benchmark; ten years later, they earn 96% of U.S.-born earnings. Filipino immigrants initially earn 59% of U.S. natives' earnings; ten years later, they earn 96%.[13,14]

---

[13] The earnings of immigrants from English-speaking Europe and Canada continue to exceed the earnings of these Asian groups.

[14] Immigrants from the developing Asian countries share a similar beginning point to the all-immigrant category (last row of Table 7.5): the median unadjusted entry earnings of all immigrants in 1979 are about 50% of U.S. natives' earnings; the median earnings of Chinese, Indians, Koreans, and Filipinos, combined, are about 55% of natives' earnings in 1979. Yet, ten years later, the Asian median earnings are 107% of natives' earnings, whereas the all-immigrant median is about 77% of natives' earnings. The common beginning point of the all-immigrant and Asian groups, yet the far higher eventual earnings of the latter, is a mani-

Table 7.5 Entry earnings and earnings 10 years later for immigrant men who entered the U.S. in 1975–1980 and were 25–54 in 1980 (median earnings, 1989 dollars) (percentage of US native born in parentheses)

	At entry: 1979 earnings (men ages 25–54)	10 years later: 1989 earnings (men ages 35–64)	10-year earnings growth rate
Philippines	15,348 (59%)	25,000 (96%)	62.9
China	9739 (37)	25,000 (96)	156.7
Korea	15,196 (58)	25,000 (96)	64.5
India	17,746 (68)	37,200 (143)	109.6
Japan	28,727 (110)	30,150 (116)	5.0
Western Europe	23,659 (90)	27,000 (104)	14.1
English speaking	32,216 (123)	44,360 (171)	37.7
Other Western Europe	20,280 (77)	26,000 (100)	28.2
Eastern Europe	13,523 (52)	27,000 (104)	99.7
Canada	30,425 (116)	38,000 (146)	24.9
All Immigrants	12,949 (49)	20,000 (77)	54.5

Source: Authors' estimates based on a 6% 1980 file created by combining and reweighting the 5% and 1% files of the 1980 Census of Population, and a 6% microdata 1990 sample created by combining and reweighting the 1990 Census of Population Public Use 5% and 1% samples. Appendix A provides information on sample sizes for entry cohorts at entry and ten years later

Notes: Native born are defined as persons born in the U.S.; foreign born are defined as persons born outside of the United States excluding those with U.S. parents. China includes mainland China, Taiwan, and Hong Kong. Korea includes South, North, and unspecified Korea. The sample does not exclude students, the self-employed, and persons with zero earnings

Table 7.6 shows by education and age the entry earnings and ten-year earnings growth of the Asian groups and a combined European/Canadian comparison group. An inverse relationship between immigrant entry earnings and earnings growth is evident. In all age–education categories,

festation of the interaction between the high education levels of Asian immigrants, relative to the all-immigrant group, and the acquisition of U.S.-specific skills. According to the IHCI Model, the lower the degree of transferability, the greater the effect of source-country human capital on the return to investment and the lesser the effect of source-country human capital upon opportunity costs. Thus, among groups lacking skill transferability, the more likely it will be that immigrants with higher levels of source-country human capital will have greater earnings growth than immigrants with lower levels of source-country human capital, and we will observe an even stronger inverse relationship between initial earnings and earnings growth. Refer to Chap. 4 on this point.

Table 7.6 Entry earnings and earnings growth of the 1975–1980 cohort by age and years of schooling, 1980 and 1990 censuses (median earnings, 1989 dollars by PCD)

	Ages 25–39; less than 13 years of schooling		Ages 25–39; 13 or more years of schooling	
	Entry earnings	Earnings growth	Entry earnings	Earnings growth
Philippines	13,523	36.8	16,902	72.3
China	10,144	47.9	8455	373.1
Korea	15,212	57.8	13,523	121.8
India	14,199	54.9	19,943	113.1
Japan	20,280	18.3	25,779	39.6
Europe/Canada	17,341	43.6	25,348	59.8
	Ages 40–54; less than 13 years of schooling		Ages 40–54; 13 or more years of schooling	
	Entry earnings	Earnings growth	Entry earnings	Earnings growth
Philippines	12,273	28.7	16,902	42.0
China	9469	5.6	14,942	33.9
Korea	15,212	31.5	16,902	25.2
India	14,250	5.3	21,970	50.2
Japan	20,280	-1.4	54,405	-43.0
Europe/Canada	17,578	10.9	33,795	-0.2

Source: Authors' estimates based on a 6% 1980 file created by combining and reweighting the 5% and 1% files of the 1980 Census of Population, and a 6% microdata 1990 sample created by combining and reweighting the 1990 Census of Population Public Use 5% and 1% samples. Appendix A provides information on sample sizes for entry cohorts at entry and ten years later

Notes: Foreign born are defined as persons born outside of the United States excluding those with U.S. parents. China includes mainland China, Taiwan, and Hong Kong. Korea includes South, North, and unspecified Korea. The sample does not exclude students, the self-employed, and persons with zero earnings

Japanese immigrant men have the highest entry earnings but lowest earnings growth; immigrant men from each of the Asian developing countries have lower entry earnings than their European/Canadian counterparts, but generally higher earnings growth.

The degree of the inverse relationship is greater for younger versus older immigrants, and for more educated versus less educated immigrants. It is clearest and most pronounced for immigrants who are both young

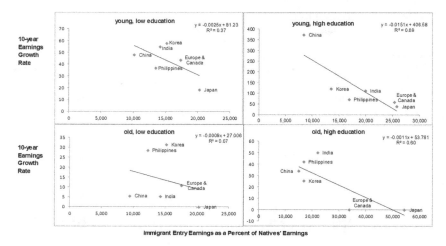

Fig. 7.2 The inverse relationship by age and education. (Source: Authors' creation based on 1980 and 1990 Census PUMS data)

and well educated. Figure 7.2, and the accompanying fitted regression lines, highlight these patterns.

Relative to U.S.-born men, Table 7.7 shows that the developing Asian country immigrants start at much lower earnings than U.S. natives do. Nevertheless, with time, their earnings grow rapidly and approach or exceed the earnings of U.S. natives of similar age and education.

Echoing a theme of this book, the entry earnings of immigrants lacking U.S.-specific human capital are poor predictors of their ultimate economic success. This is particularly pronounced for the more educated. Despite their eventual success, immigrants with some college in the Asian developing country groups start their U.S. journeys with earnings near or below that of immigrants with only a high school degree.

WHY ARE THE CROSS-SECTIONAL AND COHORT-BASED ESTIMATES OF EARNINGS GROWTH SO SIMILAR?

The cross-sectional estimates of earnings growth for Asian and European immigrants resemble the earnings growth we measure following cohorts across decennial censuses. As in the cross-sectional estimates, the earnings

Table 7.7 Entry earnings and earnings 10 years later of the 1975–1980 cohort, by age and years of schooling, relative to the native born, median earnings 1980 and 1990 censuses

	Ages 25–39 in 1980, 35–49 in 1990; less than 13 years of schooling		Ages 25–39 in 1980, 35–49 in 1990; 13 or more years of schooling	
	Entry earnings	Earnings 10 years later	Entry earnings	Earnings 10 years later
Philippines	0.57	1.09	0.53	1.14
China	0.43	0.88	0.26	1.57
Korea	0.64	1.41	0.42	1.18
India	0.60	1.29	0.62	1.67
Japan	0.86	1.41	0.80	1.41
Europe/Canada	0.73	1.46	0.79	1.59
	Ages 40–54 in 1980, 50–64 in 1990; less than 13 years of schooling		Ages 40–54 in 1980, 50–64 in 1990; 13 or more years of schooling	
	Entry earnings	Earnings 10 years later	Entry earnings	Earnings 10 years later
Philippines	0.43	1.21	0.37	0.96
China	0.33	0.77	0.33	0.80
Korea	0.54	1.54	0.37	0.85
India	0.50	1.15	0.48	1.32
Japan	0.71	1.54	1.19	1.24
Europe/Canada	0.62	1.50	0.74	1.35

Source: Authors' estimates based on a 6% 1980 file created by combining and reweighting the 5% and 1% files of the 1980 Census of Population, and a 6% microdata 1990 sample created by combining and reweighting the 1990 Census of Population Public Use 5% and 1% samples. Appendix A provides information on sample sizes for entry cohorts at entry and ten years later

Notes: Native born are defined as persons born in the U.S.; foreign born are defined as persons born outside of the United States excluding those with U.S. parents. China includes mainland China, Taiwan, and Hong Kong. Korea includes South, North, and unspecified Korea. The sample does not exclude students, the self-employed, and persons with zero earnings

growth of all the Asian groups except the Japanese exceeds that of European and Canadian immigrant men.[15]

Similarities between earnings growth estimates based on cross-sectional versus cohort analyses for post-1965 immigrants have been noted

---

[15] Note that estimates of immigrant earnings growth based on either cross-sectional data or following cohorts are affected by emigration—the topic of Chap. 10. As such, neither necessarily provides an estimate of actual earnings growth.

elsewhere leading some scholars to conclude that cross-sectional analyses of immigrant earnings are unaffected by cross-sectional bias. Yet, controlling for measured characteristics, and within many source countries,[16] we know that immigrant entry earnings declined following the 1965 Immigration and Nationality Act. How is it possible then that the cross-sectional and cohort-based estimates of earnings growth are similar?

The answer lies in the inverse relationship between entry earnings and earnings growth. If the earnings of earlier and recent immigrant cohorts converge with time in the United States, then with an over-time decline in entry earnings, the cross-sectional analysis will accurately predict the earnings profile of the most recent cohorts, but overstate the earnings growth of earlier cohorts. With an over-time increase in entry earnings, the earnings profile of the most recent cohorts will again be accurately predicted by the cross-sectional earnings profile, while the earnings growth of earlier immigrant cohorts will be underestimated. The similarity between the actual earnings growth and the cross-sectional estimates occurs because as the entry earnings of immigrants change, their earnings growth changes in the opposite direction.

Following a non-parametric approach, the cohort analyses of this chapter suggest that variations in skill transferability—not variations in immigrant quality—underlie intergroup and over time variations in immigrant earnings. The results further suggest that factors that affect immigrant entry earnings affect earnings growth. The next chapter introduces a parametric approach for measuring immigrant earnings trajectories wherein we include variables in the model that capture the effect of cohort characteristics on both entry earnings and earnings growth.

References

DeSilva, Arnold (1996) "Earnings of Immigrant Classes in the Early 1980s in Canada: A Re-examination," Working Paper, Human Resource Development Canada, 1996.

Duleep, H. and Regets, M., (1992) "Some Evidence on the Effect of Admission Criteria on Immigrant Assimilation: The Earnings Profiles of Asian Immigrants in Canada and the U.S" in *Immigration, Language and Ethnic Issues: Canada*

---

[16] Changes in the adjusted entry earnings of immigrant cohorts from the same source country could result from changes in immigrant skill transferability as a function of changes in immigrant admission restrictions and changes in the economic opportunities of the source country vis-à-vis the U.S.

and the United States, Barry Chiswick (ed.). Washington, DC: American Enterprise Institute, 1992, pp. 410–437.

Duleep, H. and Regets, M. (1996a) "Admission Criteria and Immigrant Earnings Profiles," *International Migration Review*, Summer, 30(2), 571–90.

Duleep, H. and Regets, M. (1996b) "Family Unification, Siblings, and Skills" in H. Duleep and P. V. Wunnava (editors), *Immigrants and Immigration Policy: Individual Skills, Family Ties, and Group Identities*, Greenwich, CT: JAI Press, pp. 219–244.

Friedberg, R. (1992) "The Labor Market Assimilation of Immigrants in the U.S.: The Role of Age at Arrival," Brown University.

Friedberg, R. (1993) "The Success of Young Immigrants in the U.S. Labor Market: An Evaluation of Competing Explanations," Brown University.

Jasso G, Rosenzweig MR. 1995. "Do Immigrants Screened for Skills Do Better than Family-Reunification Immigrants?" *International Migration Review* 29: 85–111.

Kossoudji, Sherrie A., "Immigrant Worker Assimilation: Is It a Labor Market Phenomenon?" *Journal of Human Resources*, Vol 24, No.3, Summer 1989, pp 494–527.

Reimers David M. 2005. *Other Immigrants: The Global Origins of the American People*, NY: NYU Press.

CHAPTER 8

Modeling the Effect of a Factor Associated with Low Entry Earnings: Family Admissions and Immigrant Earnings Profiles

A key policy issue is whether the United States should decrease family-based admissions and increase employment-based admissions.[1] Scholars attribute the age- and education-adjusted decline in immigrants' initial earnings to changes in the source-country composition of U.S. immigrants and to high family-based admissions. Chapter 7 found that country-of-origin groups with low entry earnings (conditional on their entry-levels of education and age) have high earnings growth. This chapter explores how family versus employment-based admissions affect earnings profiles.

Patterns of Admission

The post-1965 immigrants from Asian developing countries have been predominantly admitted to the United States through kinship ties: in 1980, fewer than 3% of Korean immigrants and fewer than 1% of Filipino immigrants entered via employment visas versus close to 40% of British immigrants.[2] The divergence is more striking for working-age men. Among numerically restricted immigrants (who may more accurately reflect the working population than unrestricted immigrants),

---

[1] Refer to Lowell (1996).

[2] Other European countries tend to fall in between most Asian countries and the U.K. in their admissions patterns.

60% from the United Kingdom were admitted via occupational skills, versus fewer than 2% from the Philippines and fewer than 5% from Korea. In contrast to the Asian developing countries, 61% of the numerically restricted immigrants from Japan entered via employment admissions in 1980.

Admission patterns have also varied over time. Because of discriminatory admission policies, Asian immigration to the United States was sharply curtailed from 1924 to 1965. Although the 1965 immigration reform giving preferential treatment to family members applied to all immigrants, the virtual cessation of Asian immigration for 40 years meant that for many of the first post-1965 Asian immigrants employment-based visas were the only way to gain U.S. admission; more than 60% of Indian immigrants admitted between 1965 and 1970 entered on employment visas (Fig. 8.1).

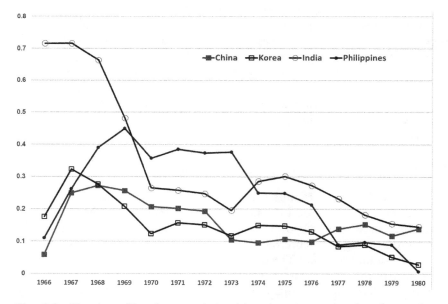

Fig. 8.1 Fraction of immigrants admitted based on occupational preference by source country. (Source: Authors' creation based on their compilation of statistics from the 1965–1980 annual reports of the Immigration and Naturalization Service)

As Asian immigrants gained a U.S. foothold, family-based admissions grew, eventually resulting in the high family admissions that continue to characterize much of the developing country Asian immigration.[3]

The opposite pattern occurred for immigrants from Western Europe and Japan: family admissions fell over time, while employment-based admissions rose. Between 1965 and 1970, fewer than 4% of British immigrants entered via employment-based visas versus 40% in 1980 (Fig. 8.2).

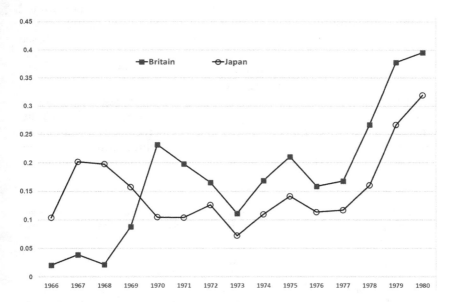

Fig. 8.2 Fraction of immigrants admitted based on occupational preference by source country. (Source: Authors' creation based on their compilation of statistics from the 1965–1980 annual reports of the Immigration and Naturalization Service)

---

[3] This analysis is based on our compilation of statistics from the 1965 to 1980 annual reports of the Immigration and Naturalization Service.

Measuring the Effect of Admission Criteria on Immigrant Earnings

Theoretically, we would expect that kinship-admitted immigrants would initially earn substantially less than those admitted via occupational skills; the very nature of employment-based admission guarantees highly transferable skills to the United States, as evidenced by an employer's willingness to participate in a cumbersome labor certification process. We would also expect family-based immigrants to experience higher earnings growth than their employment-based counterparts. Since occupation-based immigrants already have skills relevant to specific employment opportunities, their opportunity cost of investment will be greater and the return to investment in U.S.-specific human capital less than for family-based immigrants.

To test this hypothesis, we used micro data on immigrant men from the 1990 census matched to Immigration and Naturalization Service (INS) annual report data on admission criteria by year of admission and country of origin.[4] To analyze the effect of family versus occupational skills admission, we used as our explanatory variable the percentage of immigrants admitted via occupational skills in each year-of-entry/country-of-origin cohort (PerOcc).[5,6] The complement of this variable is the percentage of

---

[4] The 1990 census sample used in our analyses is a 6% sample created by combining and reweighting the Public Use 5% and 1% Public Use samples. Refer to Bureau of the Census (1992) for technical documentation on these files. We compiled the admission criteria data from the published tables of the 1975–1990 annual reports of the INS (Immigration and Naturalization Service, all years, 1975–1990).

[5] The variable, percent admitted based on occupational skills, includes both the third and the sixth occupational preference categories. It is computed as the number of immigrants admitted based on occupation skills, available by country of chargeability, divided by the total number of immigrants, available by country of birth. Thus, the first step in our data creation was to match the INS country of birth and country of chargeability records. This limited the number of countries in our data set to those for which the INS reported both country of birth and country of chargeability data and introduced a source of measurement error since individuals' country of birth does not always coincide with country of chargeability. Another source of measurement error is that the INS annual records that we used in this study do not separately report data on admissions criteria for women and men, whereas our earnings analysis is limited to men. Future analysts could use the more detailed data in INS's raw data available on year-specific tapes.

[6] The occupational-skills immigrants are those admitted under the third and sixth preferences. The preference classifications refer to those in place prior to the 1990 immigration reforms.

immigrants in a given year-of-entry/country-of-origin cohort who gained admission through family ties or refugee status.[7]

Our estimating equation is:

$$y_i = \alpha + \gamma_1 \text{PerOcc}_{jk} + X'\beta_1 + (\beta_2 + \Theta \text{PerOcc}_{jk})\text{YSM} + \varepsilon_i \qquad (8.1)$$

where y_i denotes the natural logarithm of the earnings of immigrant i; X is a vector of variables measuring age, age squared, and education (seven categories as shown in Table 8.1), and β_1 the vector of corresponding coefficients; YSM measures years since migration; and PerOcc_{jk} is the percentage of immigrants in group j and cohort k admitted on the basis of occupational skills. We would expect γ to be positive—reflecting the higher skill transferability of employment-based immigrants—and for Θ to be negative, reflecting their lower earnings growth relative to kinship-based immigrants. A negative coefficient on the YSM × PerOcc variable implies that family admissions are positively associated with immigrant earnings growth relative to occupational skills admissions. We estimated the model on three samples—all immigrants, Asian immigrants, and European immigrants. Each includes working-age men (25 to 65 years of age) who immigrated during 1975–1990 and who were at least 20 years old when they immigrated.[8,9]

The coefficients (Table 8.1) on *Percent Admitted* via *Occupation, Years since Migration*, and their interaction, indicate that as the percentage admitted via family increases, the initial earnings of immigrants fall: the

---

[7] Future work could refine the treatment of family admissions and refugee status in the model.

[8] Thus we eliminate from the analysis immigrants who entered the U.S. as children. Kossoudji (1989) suggests that with the inclusion in the study population of immigrants who migrated as children, the assimilation effect may reflect pre-labor market assimilation as opposed to labor market assimilation by immigrants. Also, see Friedberg (1992, 1993) on this issue.

[9] We coded and matched annual admission criteria data that are available in the INS annual reports for 1975–1990 to the 1990 census sample (a 6% microdata sample created by combining and reweighting the 1990 Public Use 5% and 1% Public Use samples). Europe includes 24 countries (Albania, Austria, Belgium, Bulgaria, Czechoslovakia, Denmark, Finland, France, Germany, Greece, Hungary, Ireland, Italy, Netherlands, Norway, Poland, Portugal, Romania, Spain, Sweden, Switzerland, U.K., USSR, and Yugoslavia) and Asia includes 17 countries (Burma, China, Cyprus, India, Indonesia, Iran, Iraq, Israel, Japan, Jordan, Korea, Lebanon, Pakistan, Philippines, Syria, Thailand, and Turkey).

Table 8.1 Regression of log (earnings) for immigrant men, 25–64 years old, controlling for percentage of country-of-origin/year-of-immigration cohort admitted on the basis of occupational skills (t-statistics in parentheses)

	All immigrants	Asian immigrants	European immigrants
Level of schooling			
9–11 years	0.2029 (23.37)	0.132 (5.90)	0.096 (2.90)
high school grad	0.3482 (41.68)	0.272 (13.69)	0.165 (5.83)
some college	0.4534 (55.20)	0.393 (20.64)	0.263 (9.37)
bachelor's degree	0.6688 (74.35)	0.592 (31.69)	0.490 (16.63)
master's degree	0.7463 (66.93)	0.609 (29.06)	0.586 (19.43)
professional degree	1.0096 (65.46)	1.046 (38.33)	0.720 (17.38)
PhD	0.9078 (54.52)	0.782 (28.14)	0.706 (19.44)
Age	0.0790 (33.30)	0.121 (28.63)	0.127 (19.65)
Age squared	−0.00089 (−31.86)	−0.0014 (28.73)	−0.0014 (18.39)
Percentage admitted via occupation	3.4215 (56.98)	4.488 (42.43)	2.912 (22.97)
Years since migration	0.0569 (64.20)	0.083 (44.57)	0.056 (21.73)
Years since migration x Percentage occupation	−0.1087 (−14.98)	−0.287 (22.17)	−0.147 (−9.63)
Intercept	6.9862 (148.18)	6.033 (68.99)	6.373 (47.54)
Adjusted R^2	0.25	0.22	0.20
Sample size	112, 253	42,723	18,724

Source: Authors' estimates. The 1990 census sample used in our analyses is a 6% microdata sample created by combining and reweighting the 1990 Public Use 5% and 1% Public Use samples. We matched the census data to information on admission criteria that is available in the INS annual reports for 1975–1990

Notes: The reference group for the education categorical variables is less than nine years of school. The sample excludes persons who migrated before the age of 20. Asia includes 17 countries: Burma, China, Cyprus, India, Indonesia, Iran, Iraq, Israel, Japan, Jordan, Korea, Lebanon, Pakistan, Philippines, Syria, Thailand, and Turkey. Europe includes 24 countries: Albania, Austria, Belgium, Bulgaria, Czechoslovakia, Denmark, Finland, France, Germany, Greece, Hungary, Ireland, Italy, Netherlands, Norway, Poland, Portugal, Romania, Spain, Sweden, Switzerland, the U.K., USSR, and Yugoslavia. Foreign born are defined as persons born outside of the United States excluding those with U.S. parents. The sample excludes persons who migrated before the age of 20. Zero earners are excluded because Log (Earnings) is the dependent variable. The sample does not exclude students and the self-employed

estimated coefficient on *Percent Admitted* via *Occupation* is positive. But, their expected earnings growth increases—the estimated coefficient on *Years Since Migration* × *Percent Admitted* via *Occupation* is negative.[10]

Chapter 7 revealed that immigrants from source countries with low adjusted entry earnings also had high earnings growth, a result we attribute to skill-transferability variations among immigrants from different source countries. Thus, the estimated admission criteria effects in Table 8.1 may reflect source-country variations in immigrant skill transferability. To address this issue, we include a categorical variable for each source country, alone and interacted with the set of explanatory variables X^{11}:

$$y_i = \alpha + \gamma_1 P_{jk} + \gamma_2 G_j + G_j X' \beta_1 + (\beta_2 + \Theta P_{jk}) YSM + \varepsilon_i \qquad (8.2)$$

where G_j is a categorical variable denoting group j. The revised model permits estimating separate returns to education and experience for each immigrant group, but uses the intergroup, inter-cohort variations in admission criteria to estimate γ and Θ.

---

[10] The definition of immigrant differs between INS and census data. An immigrant in the Census sample is simply a person who was born outside the U.S. who resides in the U.S. in the census year (in this analysis 1990). The INS definition of immigrant is any alien in the United States, except one legally admitted under specific nonimmigrant categories. The difference is a potential cause of measurement error in our analysis. One model would suggest that the measurement error biases the effect downward. PerOcc is the number of immigrants in a group admitted under occupational skill divided by the total number of individuals counted as "immigrants" under the INS definition. The census would count all of these people as immigrants and would count other individuals as well. Under a model where the fraction of people in the census that would not be counted by the INS is the same across groups, the coefficient on PerOcc is scaled downwards by the fraction of people in the census not counted by the INS.

[11] The country-specific dummy variables control for country-specific effects on the level of earnings (i.e. countries that tend to have high earnings may tend to have high occupational admissions as well). Interacting country-of-origin with the human capital variables in vector X controls for country-specific variations in the effects of education and other variables on earnings. This is important since, as education and age are correlated with type of admission, effects of these variables not captured by the all-country estimated effects in equation 1 will be spuriously attributed to Pjk.

Table 8.2 Regression of log (earnings) for immigrant men, 25–64 years old—the effect of percent admitted via occupational skills, controlling for source country and source country human capital interactions (t-statistics in parentheses)

	All immigrants	Asian immigrants	European immigrants
Percent admitted via occupation	2.9481 (23.89)	4.2136 (19.41)	3.0424 (11.29)
Years since migration	0.0593 (61.08)	0.0777 (34.72)	0.06618 (22.66)
Years since migration x Percent occupation	-0.1468 (-16.94)	-0.2501 (-14.27)	-0.2337 (13.55)
Adjusted R^2	0.25	0.25	0.20
Sample size	112, 253	42,723	18,724

Source: Authors' estimates. The 1990 census sample used in our analyses is a 6% microdata sample created by combining and reweighting the 1990 Public Use 5% and 1% Public Use samples. We matched the census data to information on admission criteria that is available in the INS annual records

Notes: The explanatory variables include all those that were included in the previous estimation plus dummy variables for all source regions alone and interacted with each education and age variable. Asia includes 17 countries: Burma, China, Cyprus, India, Indonesia, Iran, Iraq, Israel, Japan, Jordan, Korea, Lebanon, Pakistan, Philippines, Syria, Thailand, and Turkey. Europe includes 24 countries: Albania, Austria, Belgium, Bulgaria, Czechoslovakia, Denmark, Finland, France, Germany, Greece, Hungary, Ireland, Italy, Netherlands, Norway, Poland, Portugal, Romania, Spain, Sweden, Switzerland, the U.K., USSR, and Yugoslavia. Foreign born are defined as persons born outside of the United States excluding those with U.S. parents. The sample excludes persons who migrated before the age of 20. Zero earners are excluded because Log (Earnings) is the dependent variable. The sample does not exclude students and the self-employed

The same patterns persist. Admission based on family is associated with lower initial earnings but higher earnings growth than admission based on occupational skills (Table 8.2). For instance, if all immigrants are kinship admitted, earnings are projected to grow about 5.9% with each year since migration, holding other variables constant. If 5% are admitted with employment visas, earnings will grow at a rate of 5.2% per annum. And if 40% are employment based, as was true of British immigrants in 1980, earnings grow at about three-hundredths of a percentage per year.

EDUCATION AND THE EARNINGS' EFFECT OF ADMISSION CRITERIA

To explore how admission criteria effects vary with immigrant education level, we interacted educational achievement (Ed) with percentage admitted on the basis of occupation (PerOcc), years since migration (YSM), and the interaction between percentage admitted on the basis of occupation and years since migration (PerOcc · YSM):

$$y_i = \alpha + \gamma_1 \text{PerOcc}_{jk} + \gamma_2 \text{PerOcc}_{jk} \cdot \text{Ed} + X' \beta_1 \\ + \left(\beta_2 + \Theta_1 \text{PerOcc}_{jk} \right) \text{YSM} + \left(\beta_3 + \Theta_2 \text{PerOcc}_{jk} \right) \text{YSM} \cdot \text{Ed} + \varepsilon_i \quad (8.3)$$

Table 8.3 (second column) shows the estimated coefficients. (The first column repeats the coefficients of Model 1.) In keeping with our theoretical expectations, the interaction term of employment-based admission with years since migration remains negative in the model with the education interactions. More significantly, the lower the percentage of employment-based immigrants, the greater the extent to which higher education is associated with greater earnings growth. Since higher levels of human capital increase both the return to host country human capital investment and the opportunity cost, the IHCI Model does not predict the effect of source-country human capital on human capital investment. It does, however, predict that the harder it is to transfer skills, the more likely it is that immigrants with higher education levels will have greater rates of human capital investment.[12]

The coefficients suggest that employment-based immigrants, who have highly transferable skills, will have slower earnings growth the higher their education—18.2 percentage points less for each year of education. For family-based immigrants, the regression estimates imply that annual earnings growth is 4.3 percentage points greater for each year of education. These are unrealistically large effects obtained by setting our "percentage occupational admissions" variable to extreme values of zero and one.

---

[12] The underlying intuition is that a person with a high level of human capital in their country of origin, but low initial skill transferability, will face low opportunity costs for investment, but a high ability to gain new skills, as well as a greater incentive to gain U.S. human capital that would complement source-country skills and make them valuable in the U.S. labor market.

Table 8.3 The interactive effect of admission criteria and education on immigrant earnings: regression of log (earnings) for immigrant men, 25–64 years old, all immigrants (t-statistics in parentheses)

	Not including interaction between admission criteria and education and interaction between years since migration and education	Including interaction between admission criteria and education and interaction between years since migration and education
Percentage admitted via occupation	3.4215 (56.98)	2.9227 (34.33)
Years since migration	0.0569 (64.20)	0.0480 (48.46)
Years since migration x Percent occupation	−0.1087 (−14.98)	−0.0768 (−7.49)
Percent admitted via occupation x education	–	1.8917 (14.85)
Years since migration x education	–	0.0429 (20.48)
(years since migration x Percentage occupation) x education	–	−0.1822 (−11.48)
Adjusted R²	0.25	0.25
Sample size	112, 253	112, 253

Source: Authors' estimates. The 1990 census sample used in our analyses is a 6% microdata sample created by combining and reweighting the 1990 Public Use 5% and 1% Public Use samples. We matched the census data to information on admission criteria that is available in the INS annual reports for 1975–1990

Notes: Foreign born are defined as persons born outside of the United States excluding those with U.S. parents. The sample excludes persons who migrated before the age of 20. Zero earners are excluded because Log (Earnings) is the dependent variable. The sample does not exclude students and the self-employed

Nevertheless, both the size and the statistical significance of these coefficients strongly support the IHCI Model.

To further probe the admission criteria/education interaction, we separately estimated Model 1 on samples of immigrants divided by level of schooling (Table 8.4). Both the initial effect of admission criteria and its

Table 8.4 Regression of log (earnings) for immigrant men, 25–64 years old, controlling for percentage of country-of-origin/year-of-immigration cohort admitted via occupational skills by education level (t-statistics in parentheses)

	Bachelor's degree or higher	Less than bachelor's degree	High school diploma or higher	Less than high school diploma
All immigrants				
Percent admitted via occupation	4.7444 (48.11)	2.9835 (35.71)	4.2042 (59.37)	2.6834 (17.35)
Years since migration	0.0802 (39.63)	0.0525 (53.05)	0.0674 (53.00)	0.0481 (37.61)
Years since migration x Percent occupation	−0.2554 (−20.29)	−0.0809 (−8.04)	−0.1863 (−21.15)	−0.0298 (−1.57)
Adjusted R^2	0.21	0.15	0.20	0.08
Sample size	31,101	81,151	68,464	43,788
Asian immigrants				
Percent admitted via occupation	5.2388 (37.64)	3.7636 21.48	4.7462 (42.77)	3.1424 (7.85)
Years since migration	0.0941 (32.05)	0.0751 (30.59)	0.0883 (42.79)	0.0647 (14.09)
Years since migration x Percent occupation	−0.3089 (−17.10)	−0.2824 (−13.86)	−0.3088 (22.45)	−0.2167 (−4.83)
Adjusted R^2	0.22	0.13	0.21	0.07
Sample size	17,325	19,424	30,809	5940
European immigrants				
Percentage admitted via occupation	3.7258 (20.19)	2.3015 (13.04)	3.0319 (22.17)	2.8162 (7.21)
Years since migration	0.0608 (14.38)	0.0535 (16.43)	0.0582 (19.53)	0.0528 (9.88)
Years since migration x Percentage occupation	−0.2094 (−9.21)	−0.1042 (−4.95)	−0.1551 (9.27)	−0.1765 (3.82)
Adjusted R^2	0.19	0.10	0.17	0.06

Source: Authors' estimates. The 1990 census sample used in our analyses is a 6% microdata sample created by combining and reweighting the 1990 Public Use 5% and 1% Public Use samples. We matched the census data to information on admission criteria that is available in the INS annual records

Notes: Foreign born are defined as persons born outside of the United States excluding those with U.S. parents. The sample excludes persons who migrated before the age of 20. Zero earners are excluded because Log (Earnings) is the dependent variable. The sample does not exclude students and the self-employed

effect on earnings growth increase with level of education. In the estimations of all immigrants combined, the estimated coefficients suggest that, for those with less than a high school diploma, increasing the proportion admitted through occupational skills by one percentage point increases initial earnings by 2.68% and decreases annual earnings growth (relative to family-based admissions) by 0.03 percentage points. For those with a bachelor's degree or more, a 1% increase in occupational skills admissions is associated with a 4.74% increase in initial earnings and a 0.25 percentage point decrease in earnings growth (from a base rate of 8.02 percent) with years spent in the United States. This is generally consistent with the regression results in Table 8.3 that included interactions with years of education.

For all but one high/low education comparison, higher levels of education are associated with lower earnings growth for the employment-based immigrants (evaluated using the coefficients of both the years-since-migration variable and its interaction with percentage admitted through occupation). The exception, when European immigrants are split between high-school completion and less than high school, may be the result of a generational effect in high-school completion for Europeans; older workers have less incentive to invest in new labor market skills. However, when Europeans are split by college completion, the decrease in earnings growth with higher education for employment admissions is large.

Concluding Remarks

Greater earnings growth for family-based admissions, controlling for age and education, is consistent with less transferable human capital. Similarly, the greater inverse relationship between initial earnings and earnings growth for Asian than for European immigrants is consistent with the IHCI model if the skill transferability of Asian immigrants is lower than the skill transferability of European immigrants. That earnings growth increases with education for family-based immigrants, but falls with education for employment-based immigrants, exactly conforms to the IHCI Model's predictions.

REFERENCES

Friedberg, R. (1992) "The Labor Market Assimilation of Immigrants in the U.S.: The Role of Age at Arrival," Brown University.

Friedberg, R. (1993) "The Success of Young Immigrants in the U.S. Labor Market: An Evaluation of Competing Explanations," Brown University.

Kossoudji, Sherrie A., "Immigrant Worker Assimilation: Is It a Labor Market Phenomenon?" *Journal of Human Resources*, Vol 24, No.3, Summer 1989, pp. 494–527.

Lowell, B. Lindsay (1996) "Skilled and Family-Based Immigration: Principles and Labor Markets," in *Immigrants and Immigration Policy: Individual Skills, Family Ties, and Group Identities*, Greenwich, CT: JAI Press.

CHAPTER 9

Human Capital Investment

Human capital investment takes myriad forms some of which are difficult even impossible to measure. Failing to measure human capital investment for immigrants with high earnings growth does not imply the absence of human capital investment. With this caveat in mind, we examine three forms of human capital investment—learning English, changing jobs, and attending school.

INVESTMENT IN ENGLISH PROFICIENCY

English proficiency increases the transferability of source-country skills and facilitates the acquisition and use in the labor market of U.S. skills. To measure the English investment of the cohort of immigrant men who entered the United States in 1975–1980, we measure their initial English proficiency, with the 1980 census, and their proficiency ten years later, with the 1990 census. Given the near universality of English in Canada, the UK, and Ireland, a comparison of English investment between immigrants from these source regions and Asian immigrants is meaningless.[1] We instead compare the linguistic progress of the developing-country Asian groups with immigrants from the non-English-speaking countries of Western Europe.

---

[1] English proficiency investment would necessarily be greater for the Asian immigrants.

When we measure English investment by the percentage of immigrants who speak English very well, the evidence is unequivocal (Table 9.1). From 1980 to 1990, the percentage of proficient English speakers in each of the Asian developing-country groups exceeds the corresponding gain for West Europeans. In absolute terms, English proficiency increased most for Filipino men (a gain of 13.2 percentage points); Chinese and Korean immigrant men had the highest relative gains in the percentage of English-proficient speakers.

Measuring English investment by the percentage of immigrants who speak English poorly, the evidence is mixed. The percentage change for Chinese immigrants is slightly lower than it is for the non-English-speaking West European group. The varied results may be because some groups are more likely to live and work in enclaves. Chiswick and Miller (1992) find that living in a language enclave is negatively associated with acquisition of the host-country language, especially for poorly educated immigrants. Case studies suggest that enclaves help immigrants lacking access to primary sector jobs bypass the confines of the secondary sector.[2] McManus (1990) finds that large enclaves provide better jobs for persons who are not English proficient. Bauer, Epstein, and Gang (2005) find that Mexican migrants with poor English skills migrate to locations with a large enclave.

Particularly striking is the reduction in poor English speakers for East Europeans. They start with a higher percentage of persons with poor English proficiency than any of the Asian developing country groups yet experience a much greater reduction, 44% to 18%, versus 39% to 31% for the Chinese and 43% to 30% for the Koreans. A lesser presence of enclaves among the East Europeans, which may reflect a greater presence of refugees,[3] could contribute to these findings.[4]

---

[2] See for instance Jiobu (1996), Gallo and Bailey (1996), Bailey (1987), Bailey and Waldinger (1991), Waldinger (1986, 1989), Portes and Bach (1985), Light (1972, 1984), Light et al. (1994), Logan, Zhang and Alba (2002), Kaplan (1997), and Wilson and Portes (1980).

[3] The Asian groups that we study in Parts II through IV include few refugees. In contrast, during the Cold War, and up until the mid-1990s, many U.S. refugees came from the former Soviet Union. Two factors work against enclave creation for refugees. One is a U.S. policy of geographically dispersing refugees. The other is that refugee migration often arises from a sudden change.

[4] Refer to Cortes (2004).

Table 9.1 Ability to speak English of the 1975–1980 immigrant entry cohort measured in 1980 and 1990

	Speaks English poorly or not at all (percent)					Speaks only English or speaks English very well (percent)					
	1980, ages 25–54	1990, ages 35–64	Percentage point change	Relative to W. Europeans	Percent change	Relative to W. Europeans	1980 ages 25–54	1990 ages 35–64	Percentage point change	Relative to W. Europeans	Percent change	Relative to W. Europeans
Philippines	6.3	3.6	-2.7	0.40	-43%	1.87	54.5	67.7	13.2	2.44	24%	1.85
China	39.0	30.9	-8.1	1.19	-21%	0.91	20.1	30.5	10.4	1.93	52%	4.00
Korea	42.9	30.4	-12.5	1.84	-29%	1.26	15.9	25.1	9.2	1.70	58%	4.46
India	5.8	3.1	-2.7	0.40	-47%	2.04	68.0	80.1	12.1	2.24	18%	1.38
Japan	26.6	17.8	-8.8	1.29	-33%	1.43	25.5	41.9	16.4	3.04	64%	4.92
Non-English-speaking Western Europe	29.3	22.5	-6.8	1.00	-23%	1.00	42.2	47.6	5.4	1.00	13%	1.00
Eastern Europe	43.9	18.1	-25.8	3.79	-59%	2.57	23.1	45.6	22.5	4.17	97%	7.46

Source: Authors' estimates based on a 6% 1980 file created by combining and reweighting the 5% and 1% files of the 1980 Census of Population and a 6% microdata sample created by combining and reweighting the 1990 Census of Population Public Use 5% and 1% Public Use samples. Appendix A provides information on sample sizes by entry cohort

Notes: Foreign born are defined as persons born outside of the United States excluding those with U.S. parents. The sample does not exclude students, the self-employed, and persons with zero earnings

Occupational Change with Time in the United States

Changing jobs is a form of human capital investment and an indicator that human capital investment has occurred. Given their likely lower skills transferability, we would expect more job changes for Asian developing-country immigrants than for European immigrant men with comparable levels of education and experience.

To test this prediction, we separately estimate a multinomial logit model of occupational choice for Asian and European immigrant men controlling for education, experience (age minus years of schooling minus 6), year of immigration, and English proficiency. ("Laborer" serves as the reference group.) Our estimation measures the effect of recent immigration on the probability of employment in different occupations. Although we don't observe occupational change for individual immigrants, we compare the occupation of more recent immigrants with that of similar earlier immigrants.[5]

Skill transferability differences suggest that being a recent immigrant (an immigrant who has been in the United States for five or fewer years), would profoundly affect occupation for the developing-country Asian immigrant men while having a smaller effect for European immigrants. To facilitate interpretation of the logit coefficients, we estimate the probability of being in each occupational class, versus being a laborer, for immigrants who have 10 years of labor market experience, 12 years of schooling, and who speak English well, first assuming U.S. residence for five or fewer years, and then assuming U.S. residence for more than 5 years (Table 9.2).

For Asians, recent immigration decreases the chances of white-collar employment by 10 to 22 percentage points, with the chances of entering the professional class the most affected. No statistically significant effects are measured for Europeans. The contrast suggests that Asian immigrant men change their occupation with time in the United States, whereas similarly educated and experienced European immigrant men are less likely to do so.

Green (1999) finds a greater propensity of immigrants to change occupations than natives beyond what can be explained by an assimilation effect. Green's result combined with the IHCI model suggests that low-skill transferability immigrants, such as family-based immigrants and

---

[5] Further work is needed to see if these cross-sectional results hold with longitudinal data or with following cohorts across cross-sections.

Table 9.2 Effect of recent immigration on occupation, immigrant men Ages 25–64

Occupational group	Asian immigrants The decrease in probability with recent immigration of being in occupation *i* versus laborer	European immigrants The decrease in probability with recent immigration of being in occupation *i* versus laborer
Manager	-0.16*	-0.03
Professional	-0.22*	-0.06
Technical	-0.10*	-0.01
Sales	-0.13*	-0.03
Clerical	-0.01	-0.06
Craft	-0.06*	-0.02
Operative	0.02	-0.03
Farm/Fish	-0.05	0.004
Service	-0.08*	-0.03
Domestic	0.0001	–

Source: Authors' estimates based on a 6% 1980 file created by combining and reweighting the 5% and 1% files of the 1980 Census of Population

Notes: Foreign born are defined as persons born outside of the United States excluding those with U.S. parents. The sample does not exclude students, the self-employed, and persons with zero earnings.
*Statistically significant at 0.05 level

immigrants from economically developing countries, provide the United States a flexible source of human capital that responds to the ever-changing needs of a dynamic economy. When demand shifts, requiring new skills to be learned, immigrants who lack host-country-specific skills will be more likely to pursue new opportunities than will natives or immigrants with highly transferable skills.

INVESTMENT IN SCHOOLING

Because of their lower opportunity costs and the usefulness of non-transferable human capital for learning, immigrants who initially lack U.S.-specific skills may undertake greater human capital investment beyond what enables the transference of previously learned skills. We would therefore expect greater investment in schooling—and at older ages—by immigrant men from the Asian developing countries than by immigrants from Western Europe and Canada, or by U.S. natives.

Table 9.3 (first column) shows the group percentages of immigrant men who are 1975–1980 entrants and ages 25–54 in 1980 who reported attending school in 1980. School attendance for Korean and Indian men is nearly double, and for the Chinese triple, the West European benchmark. On the other hand, Filipino men are somewhat less likely to attend school than the West Europeans. In 1990, 6.9% of U.S.-born men, ages 25–54, attended school. This is roughly half the corresponding percentage for all foreign-born men (13.1%).

We also examine school attendance for the 1975–1980 cohort ten years later, when immigrants are 35 to 64 years old (Table 9.3, third column). Although school attendance decreases with age, the relatively greater propensity of the developing-country Asian immigrants to attend school persists. For all of the Asian immigrant groups (including the Filipinos), school attendance is nearly double the corresponding West European percentage.

Table 9.3 School attendance of the 1975–1980 immigrant entry cohort measure in 1980 and 1990

Attended school in...				
	1980, ages 25–54	Relative to Western Europe	1990, ages 35–64	Relative to Western Europe
Philippines	8.9	0.99	5.9	2.03
China	30.4	3.37	5.9	2.03
Korea	17.7	1.97	8.2	2.83
India	17.9	1.99	5.4	1.86
Japan	16.8	1.87	5.7	1.97
Western Europe	9.0	1.00	2.9	1.00
Eastern Europe	8.8	0.98	4.3	1.48
Canada	11.7	1.30	2.2	0.78

Source: Authors' estimates based on a 6% 1980 file created by combining and reweighting the 5% and 1% files of the 1980 Census of Population and a 6% microdata sample created by combining and reweighting the 1990 Census of Population Public Use 5% and 1% Public Use samples. Appendix A provides information on sample sizes by entry cohort

Notes: Foreign born are defined as persons born outside of the United States excluding those with U.S. parents. The sample does not exclude students, the self-employed, and persons with zero earnings

Table 9.4 follows schooling achievement.[6] The left-hand side tracks from 1980 to 1990 the percentage of 1975–80 entrants lacking a high school degree. For men from the Philippines, China, Korea, and India, the percentage of men lacking a high school degree decreases more than the decrease for West European men. Particularly impressive is the growth in high school completion for Filipino men. The right-hand side tracks the percentage who report some college. It shows larger increases for all of the Asian developing-country groups than for the West Europeans, with an especially large increase for Filipino men.

The Japanese results show that the percentage of men with less than a high school degree *increases* while the percentage with some college *decreases*. Emigration clearly affects these results. Among other things (discussed in the next chapter), the findings of Table 9.4 suggest that those who return to Japan have higher initial schooling levels than those who stay in the United States.

We do not know why Filipino men show high educational investment with their ten-year change in school achievement but not with the school attendance measure. Given their low emigration rates (Chap. 10) it is unlikely that the difference reflects emigration. Using the New Immigrant Survey to examine different types of human capital investment by immigrants, Akresh (2007) finds that, even while enrolled in school, most immigrants continue to work full time or close to full time. It may be that, for some groups, the census school attendance question does not effectively measure part-time schooling.

Concluding Remarks

The steps that pave the path of economic assimilation are numerous, varied, and hard to measure. The form that human capital investment takes will depend on context, history, and community.[7] Chapter 4 argued that

---

[6] Doing this is problematic since the method by which the census measures schooling achievement changed between 1980 and 1990. The 1980 census measured schooling achievement by years completed; the 1990 census measures the attainment of certain levels of schooling, such as the receipt of a high school diploma. Furthermore, such comparisons have to be limited to comparisons above or below a certain point. Otherwise, one is faced with the difficulty of interpreting situations such as a simultaneous decrease in those with a college degree and an increase in those with a graduate degree.

[7] There is a large literature by sociologists and historians concerning what leads certain groups to follow certain strategies. See, for instance, Portes (1995a, b), Portes and Bach

Table 9.4 Educational level of the 1975–1980 immigrant entry cohort measured in 1980 and 1990

	Percent with less than a high school degree				Percent with some college				
	1980, ages 25–54	1990, ages 35–64	Percentage point change	Percent change	Change in percent relative to Western Europe	1980, ages 25–54	1990, ages 35–64	Percentage point change	Percent change	Change in percent relative to Western Europe
Philippines	18.1	6.9	-11.2	-62%	37.30	68.6	79.0	10.4	15%	2.66
China	26.9	24.6	-2.3	-9%	7.67	57.3	61.5	4.2	7%	1.08
Korea	11.1	9.8	-1.3	-12%	4.33	62.4	70.3	7.9	13%	2.03
India	10.8	6.9	-3.9	-36%	13.00	79.6	85.8	6.2	8%	1.59
Japan	3.7	7.4	3.7	100%	–	80.9	70.2	-10.7	-13%	–
Western Europe	27.0	26.7	-0.3	-1%	1.00	52.8	56.7	3.9	7%	1.00
Eastern Europe	27.0	22.0	-5.0	-18%	16.77	52.2	51.5	-0.7	-1%	–
Canada	16.4	15.2	-1.2	-7%	4.00	64.4	68.9	4.5	7%	1.15

Source: Authors' estimates based on a 6% 1980 file created by combining and reweighting the 5% and 1% files of the 1980 Census of Population and a 6% microdata sample created by combining and reweighting the 1990 Census of Population Public Use 5% and 1% Public Use samples. Appendix A provides information on sample sizes for entry cohorts at entry and ten years later

Notes: Foreign born are defined as persons born outside of the United States excluding those with U.S. parents. The sample does not exclude students, the self-employed, and persons with zero earnings. The educational categories on the 1980 and 1990 census are not completely comparable; the 1980 schooling is in terms of years whereas the 1990 schooling is in terms of levels

the adjusted earnings' gap (the difference in earnings between immigrants and natives with comparable levels of education) embodies all aspects of skill transferability, measured and unmeasured. Similarly, earnings growth embodies all aspects of human capital investment, measured and unmeasured. Nevertheless, this chapter's analysis of three forms of human capital investment is generally consistent with the IHCI model: Asian immigrant men from developing countries tend to have higher rates of investment in English proficiency, occupational change, and schooling than immigrants from Western Europe.

The school attendance results are particularly important. Ten-year changes across decennial censuses in English proficiency, occupation, and education reflect, to an unknown degree, emigration and other intercensus changes such as how an immigrant group identifies itself. School attendance is not afflicted by inter-census changes because it measures current human capital investment. In considering immigration's impact on the U.S. economy, an important issue is whether census-measured schooling investment primarily represents learning English. Since the census school attendance question asks only about courses in degree or high school diploma programs, it is not measuring English acquisition per se.

Immigrants who cannot initially earn at the level of natives with similar education levels benefit their host countries' economies by providing a source of human capital that can be applied to the learning of new skills. The higher propensity to invest in human capital, beyond that which restores the value of immigrants' original human capital—as exemplified by the school attendance statistics—makes low-skill-transferability immigrants a flexible and dynamic component of the U.S. labor force.

The fairly high Japanese result on attending school highlights a distinction for future research. The educational investment may reflect immigrants pursuing U.S. schooling and then returning to their home countries: the more transferable immigrant skills are to the U.S., the more transferable U.S.-acquired skills will be to the home country and the higher the propensity to emigrate. Furthermore, the more similar (or favorable) the

(1985), Portes and Jensen (1987), Portes and Mozo (1985), Portes and Rumbaut (1996), Portes and Stepick (1985), and Portes and Truelove (1987), Portes and Zhou (1993) and D. Reimers (2005).

economic opportunities are of the home country relative to the United States, the more likely that emigration will take place. By inverting the two causes of intergroup differences in skill transferability discussed in Chap. 3, we would expect that among immigrants who attend school in the United States, immigrants from economically developed countries would be more likely to return home after schooling than would immigrants from less economically developed countries.

References

Akresh IR. "U.S. immigrants' labor market adjustment: additional human capital investment and earnings growth." *Demography*. 2007;44(4):865–881.

Bailey, Thomas R. (1987) *Immigrant and Native Workers: Contrasts and Competition*, Conservation of Human Resources, Boulder and London: Westview Press.

Bailey, Thomas R. and Roger Waldinger (1991) "Primary, Secondary, and Enclave Labor Markets: A Training Systems Approach," *American Sociological Review*, vol. 56, August, pp. 432–445.

Bauer, Thomas, Gil S. Epstein, and Ira N. Gang (2005) "Enclaves, Language and the Location Choice of Migrants," *Journal of Population Economics*, 18 (4), 649–662

Chiswick, B. R. and Miller, P.W. (1992) "Language in the Immigrant Labor Market" in ed. Barry Chiswick, *Immigration, Language, and Ethnicity: Canada and the United States*, Washington, DC: American Enterprise Institute, pp. 229–296

Cortes, K. E. (2004) "Are Refugees Different from Economic Immigrants? Some Empirical Evidence on the Heterogeneity of Immigrant Groups in the United States," *Review of Economics and Statistics*, 2004, 86 (2), 465–480

Gallo C, Bailey TR. (1996) "Social Networks and Skills-Based Immigration Policy," in *Immigrants and Immigration Policy: Individual Skills, Family Ties, and Group Identities*, HO Duleep, PV Wunnava, eds. Greenwich, Conn.: JAI Press.

Green DA. 1999. "Immigrant Occupational Attainment: Assimilation and Mobility Over Time." *Journal of Labor Economics* 17: 49–79.

Jiobu RM. 1996. "Explaining the Ethnic Effect," in *Immigrants and Immigration Policy: Individual Skills, Family Ties, and Group Identities*, HO Duleep, PV Wunnava, eds. Greenwich, Conn.: JAI Press.

Kaplan, D.H. (1997) "The Creation of an Ethnic Economy: Indochinese Business Expansion in Saint Paul." *Economic Geography* 73(2):214–233.

Light, I. (1984) "Immigrant and Ethnic Enterprise in North America." *Ethnic and Racial Studies* 7:195-216.

Light, I. (1972) *Ethnic Enterprises in America: Business and Welfare Among Chinese, Japanese, and Blacks.* Berkeley, Calif: University of California Press.

Light, Ivan, Georges Sabagh, Mehdi Bozorgmehr and Claudia Der-Martirosian (1994) "Beyond the Ethnic Enclave Economy," *Social Problems*, Vol. 41, No. 1, Special Issue on Immigration, Race, and Ethnicity in America (Feb.), pp. 65–80.

Logan, John R., Wenquan Zhang and Richard D. Alba (2002) "Immigrant Enclaves and Ethnic Communities in New York and Los Angeles," *American Sociological Review*, Vol. 67, No. 2 (Apr.,), pp. 299–322

McManus, Walter S. (1990): "Labor Market Effects of Language Enclaves: Hispanic Men in the United States," *Journal of Human Resources*, 25(2), 228–252.

Portes A. 1995a. "Economic Sociology and the Sociology of Immigration: A Conceptual Overview," in *The Economic Sociology of Immigration: Essays on Networks, Ethnicity, and Entrepreneurship*, A Portes, ed. New York: Russell Sage Foundation.

Portes A. 1995b. "Children of Immigrants: Segmented Assimilation and Its Determinants," in *The Economic Sociology of Immigration: Essays on Networks, Ethnicity, and Entrepreneurship*, A Portes, ed. New York: Russell Sage Foundation.

Portes A and Bach R. (1985) *Latin Journey: Cuban and Mexican Immigrants in the United States.* Berkeley, Calif.: University of California Press.

Portes, Alejandro and Leif Jensen, "What's an Ethnic Enclave? The Case for Conceptual Clarity," *American Sociological Review*, vol. 52, December, 1987, pp. 768–771.

Portes, Alejandro and Rafael Mozo. (1985) "The Political Adaptation Process of Cubans and Other Ethnic Minorities in the United States: A Preliminary Analysis," *International Migration Review*, vol. 19, no. 1 Spring.

Portes A and Rumbaut RG. (1996) *Immigrant America: A Portrait* (2nd ed.). Berkeley, Calif.: University of California Press.

Portes, A. and Stepick, A. (1985) "Unwelcome Immigrants: The Labor Market Experiences of 1980 (Mariel) Cuban and Haitian Refugees in South Florida," *American Sociological Review* 50(4):493–514.

Portes, A. and Truelove, C. (1987) "Making Sense of Diversity: Recent Research on Hispanic Minorities in the United States," *Annual Review of Sociology* 13:359–85

Portes, Alejandro and Min Zhou. 1993. "The New Second Generation: Segmented Assimilation and Its Variants." *Annals of the American Academy of Political and Social Sciences* 530: 74–96.

Reimers David M. 2005. *Other Immigrants: The Global Origins of the American People*. NY: NYU Press.

Waldinger R. (1986) *Through the Eye of the Needle: Immigrant Enterprise in New York's Garment Trades*. New York: New York University Press.

Waldinger R. (1989) "Structural Opportunity or Ethnic Advantage? Immigrant Business Development in New York." *International Migration Review* 23: 48–72.

Wilson, Kenneth L. and Alejandro Portes (1980) "Immigrant Enclaves: An Analysis of the Labor Market Experiences of Cubans in Miami,"*American Journal of Sociology*, Vol. 86, No. 2 (Sept.) pp. 295–319

CHAPTER 10

Permanence and the Propensity to Invest

Starting a business, pursuing jobs with on-the-job training, and learning English take time and money and generally result in lower earnings at first. Immigrants would embark on these pursuits only if the benefits from making them could be reaped in the future.[1] This suggests that the decision to invest in U.S. human capital and the decision to stay in the U.S. are jointly determined.

If the circumstances of migration jointly determine the propensity to invest in U.S. human capital and the propensity to stay, then there should be a positive correlation between the expected return to investment in U.S. human capital that immigrants initially face and the degree to which they subsequently stay. Immigrants who do not intend on staying are likely persons who can work in the United States without investing in U.S.-specific human capital (employees of foreign-owned firms and persons who come to work in U.S. jobs requiring minimal U.S.-specific skills) or persons whose source-country backgrounds so resemble U.S. backgrounds as to obviate the need to invest in U.S.-specific skills; Canadian and British immigrants should have high emigration rates.

---

[1] The potential importance of permanence as a factor affecting immigrant investment has been discussed and explored in a variety of contexts (e.g. Erikson 1972; Piore 1979; Portes and Bach 1985; Duleep 1988; Duleep and Regets 1999; Duleep and Sanders 1993; Chiswick and Miller 1998).

Of course, individuals who initially plan on returning to their home country may end up making the U.S. their permanent home. And, the very act of investing in human capital encourages staying in the host country. Nevertheless, initial intentions are likely reflected in the human capital investment, hence earnings growth, of immigrants.

How can we measure permanence? Studies that measure naturalization rates by year of entry from the census or other cross-sectional data inaccurately gauge permanence because they ignore the increasingly select sample that occurs with emigration; a group that has a high emigration rate may appear to have a high propensity to stay. Alejandro Portes and Rafael Mozo (1985) introduced a valid way to measure intergroup variations in permanence: calculate from immigration and naturalization records the percentage of *all* entering immigrants in a particular year who subsequently become citizens. According to the Portes/Mozo measure, among post-1965 immigrants, Asian immigrants are far more likely to view the United States as their permanent home than are West Europeans and Canadians. For immigrants who entered the United States in 1971, 55% of the Asians had naturalized by 1980 versus 16% of West Europeans and less than 7% of Canadian immigrants (Table 10.1).

Researchers can also measure permanence by emigration. Using the 1980 and 1990 censuses, we follow immigrant men who entered the United States between 1975 and 1980. Their emigration rates, shown in

Table 10.1 Naturalization rates of 1971 cohort of immigrants from Asia, Western Europe, and Canada

Cumulative percentage of cohort that is naturalized by year

	1971	1972	1973	1974	1975	1976	1977	1978	1979	1980
Asia	0.40	1.62	2.31	3.58	7.73	11.22	31.44	45.03	51.13	55.31
Western Europe (including UK)	0.05	0.07	0.18	0.48	1.64	2.75	7.76	12.18	14.88	16.55
Canada	0.17	0.29	0.38	0.67	1.66	2.60	3.99	5.13	5.88	6.79

Source: Based on statistics from Immigration and Naturalization Service Annual Reports. Presented in Alejandro Portes and Rafael Mozo, "The Political Adaptation Process of Cubans and Other Ethnic Minorities in the U.S.: A Preliminary Analysis," *International Migration Review*, vol. 19, no.1, Spring 1985

Note: Although there is a minimum waiting period for naturalization (usually five years or two years for those marrying American citizens), naturalizations that appear to occur earlier than this are likely due to misreporting of the U.S. entry date

Table 10.2 Estimated change in cohort size between 1980 and 1990 and entry earnings: 25–54-year-old men in 1980 who immigrated in 1975–1980

	1990 cohort size minus 1980 cohort size (percent change)	1979 earnings of 1975–80 cohort as a percent of US-born earnings
Filipino	5.28	58.6
Chinese	-7.37	37.2
Korean	-14.38	58.0
Indian	-7.88	67.8
Filipino, Chinese, Korean, and Indian combined	-5.93	58.6
Japanese	-73.94	116.0
Western Europe	-12.67	90.4
English speaking	-28.00	123.0
Non-English speaking	-5.48	77.5
Eastern Europe	-59.83	51.6
Canada	-36.75	116.2

Source: Authors' estimates based on a 6% 1980 file created by combining and reweighting the 5% and 1% files of the 1980 Census of Population and a 6% microdata 1990 sample created by combining and reweighting the 1990 Census of Population Public Use 5% and 1% samples. Appendix A provides information on sample sizes by entry cohort

Notes: Foreign born are defined as persons born outside of the United States excluding those with U.S. parents. China includes mainland China, Taiwan, and Hong Kong. Korea includes South, North, and unspecified Korea. The sample does not exclude students, the self-employed, and persons with zero earnings

the first column of Table 10.2, are simply the percentage differences between the census samples in the number of 1975–1980 cohort immigrants.[2] Filipino men rank first in permanence. The Japanese rank last; 74% of the 1975–1980 entrants had left by 1990. Among our comparison groups, immigrants from English-speaking Europe and especially Canada have high emigration rates, consonant with the idea that immigrants who do not plan on staying in the United States are likely to be persons with skills that are highly transferable to the United States.[3]

---

[2] Although this simple emigration measure ignores important factors such as mortality, it does provide a rough measure of relative permanence. For a more sophisticated treatment of emigration for all countries, refer to Ahmed and Robinson (1994).

[3] The surprisingly high emigration rate that we measure for immigrants from the former Soviet Union countries may reflect an increasing reluctance with time in the U.S. among immigrants from these countries to identify themselves on the census as being from these countries. Mortality differences reflecting an older age structure is another probable cause. Future analyses of emigration rates should take into account mortality differences.

The second column of Table 10.2 shows for each group the median entry earnings of immigrant men as a percentage of U.S.-born median earnings. We hypothesize that the closer immigrants' entry earnings are to U.S. natives' earnings, the higher immigrant skill transferability, the lower the return to investment in U.S.-specific human capital, and the lower the propensity to stay. Comparing the first and second columns reveals a rough correspondence between this measure of the return to investment in U.S.-specific human capital and the propensity to leave. Those groups with median entry earnings that exceed U.S. natives' earnings have high emigration rates; those with relatively low entry earnings have low emigration rates.[4]

The return-to-investment measure shown in Table 10.2 ignores relevant individual characteristics such as age, schooling level, and location. A better measure is the difference between immigrant entry earnings and the earnings of U.S.-born men with similar characteristics and general skills levels, or the earnings difference between recently arrived immigrants and immigrants of the same group who share similar characteristics and general skill levels, but who have lived many years in the United States.

For each group of immigrants who entered the United States in 1975 to 1980, Fig. 10.1 graphs the potential return to investment in U.S.-specific human capital (shown on the y axis) against their probability of staying from 1980 to 1990 (shown on the x axis). The probability of staying, computed from the group-specific emigration rates, is 100 minus the percentage of men in the 1975–1980 cohort who emigrated between 1980 and 1990.[5] The potential return to investment is the earnings difference between recently arrived immigrated men and immigrant men in the same source-country group who have been in the United States 30 or more years, holding all other characteristics constant.[6]

---

[4] The exception to this pattern are immigrants from the former Soviet Union countries, whose entry earnings relative to U.S. natives are low, and who also have, by our measure, relatively high emigration. Note 3 describes a key deficiency in our emigration rate calculations.

[5] We further scaled the probability-of-staying estimates to a 0–100 scale by dividing all of the group-estimated probabilities by the computed Filipino probability of staying (105.28) and multiplying by 100.

[6] Immigrant men in the same source-country group who have been in the U.S. 30 or more years likely will have a high representation of persons who came to the U.S. as children. We of course don't know whether individuals over age 20 migrated as children because the 1980 census only records "before 1950" as the earliest period of immigration. However, an analysis of the fraction of individuals speaking only English or speaking it well by age group sug-

10 PERMANENCE AND THE PROPENSITY TO INVEST 111

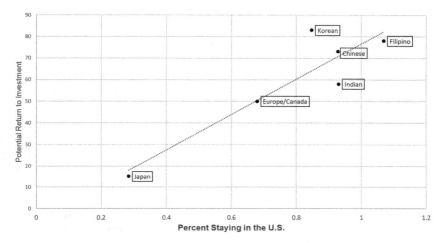

Fig. 10.1 Permanence and the potential return to investment. (Source: Authors' creation)

The points plot a positive relationship between the return to investment and the propensity to stay. At one end are Filipino immigrants, whose high-expected return to investment matches their exceptionally high degree of permanence; at the other end are the Japanese, whose low expected return to investment matches their low degree of permanence. Korean immigrants are somewhat of an outlier. Given their very large potential return to investment in U.S.-specific human capital, their permanence is less than we would expect, and may reflect the extraordinary growth of the Korean economy during this period: between 1980 and 1990, Korea's gross domestic product increased 91.4%, compared with 12.9% for China, 21.0% for India, and −20.5% for the Philippines (Heston and Summers 1996).[7]

gests that many of the migrants who came before 1950 came as children. These individuals strike us as good comparisons as they are from the same ethnic background but since they migrated early had many of the advantages of native born.

[7] The negative growth in GDP for the Philippines may also contribute to the extremely high level of permanence for Filipino immigrant men.

Japanese Immigrants and Permanence

Unlike Asian immigrants from economically developing countries, recently arrived Japanese immigrant men do not earn significantly less than comparable long-term Japanese immigrants. Despite extremely low English proficiency, their entry earnings resemble those of British and Canadian immigrants.

To unravel this puzzle, recall the discussion in Chap. 3 about the origins of skill transferability differences among immigrants. Such differences may arise from differences in the language, education, and work systems of the home and host countries, or from the selection of immigrants given differential opportunities and conditions in the home versus host country. With the generally favorable job opportunities and living conditions in Japan vis-à-vis the United States, Japanese men would be reluctant to come to the United States if their migration required investment in skills that were not easily transferable back to Japan. Of those who come, most would be persons who could work in the United States without investing in U.S.-specific skills. The likely explanation for their high entry earnings paired with low English proficiency is that many work in Japanese firms located in the United States or in firms with Japanese interests that utilize their country-of-origin training and background. Several findings support this conclusion.

The occupational distribution of Japanese immigrant men is dominated by managers, professionals, and technical workers; in 1980, 54% of Japanese immigrant men reported these occupations compared with 26% of U.S.-born non-Hispanic white men. While the propensity to become a U.S. citizen is generally high for Asian immigrants, it is low for Japanese immigrants, and particularly low in certain occupations; 86% of Japanese managers in the United States retain their Japanese citizenship. Japanese immigrant managers who are Japanese citizens earned $33,582 on average in 1980 whereas those who were U.S. citizens earned $22,372, despite the former group being four times less likely to report speaking English well than the latter group. In general, Japanese immigrant men who speak English poorly earn more than those who report speaking English well.

These findings suggest that the majority of Japanese immigrant men suffer no earnings loss because the skills they acquired in Japan are immediately relevant to their U.S. pursuits. Nor would they undertake U.S.-specific investments that would only result in higher earnings with permanent U.S. residence, since their intention is to return to Japan. Only

Table 10.3 Percentage effect of years since migration on annual earnings of foreign-born Japanese men, ages 25–64, 1980 (Benchmark group is foreign-born Japanese men who immigrated before 1950)

	Estimated effect on annual earnings (T-statistics in parentheses)	
Year of immigration	All	U.S. citizens only
1975–1980	0.032 (0.25)	-0.541 (2.41)**
1970–1974	-0.096 (0.75)	-0.336 (1.74)*
1965–1969	0.016 (0.12)	0.113 (0.65)
1960–1964	-0.027 (0.21)	-0.126 (0.76)
1950–1959	-0.064 (0.54)	-0.127 (0.93)

Source: Authors' estimates based on the 1980 Census of Population, 5% "A" Public Use Sample
*Significant at 0.10 level; **Significant at 0.05 level

a minority would plan to stay permanently in the United States and invest in U.S.-specific human capital.

To pursue this issue further, we estimated the effect of years since migration, controlling for all other observable relevant variables, on the small subset of Japanese immigrant men who had become U.S. citizens by the year 1980. Excluding immigrants who have retained their Japanese citizenship eliminates from the sample people who intend to return to Japan. The second column of Table 10.3 shows—for U.S. citizens only—the estimated percentage differences between the earnings of immigrant men in each cohort and the earnings of immigrant men who arrived 30 years ago or more, net of differences in skills and characteristics. The coefficient in the first line shows the estimated effect on earnings of having been in the United States 5 years or less versus 30 years or more. The same analysis, but estimated across all U.S. Japanese immigrants, citizens and non-citizens, is shown in the first column.

In sharp contrast with the full sample results, recent immigrants among Japanese immigrant men who clearly intend to stay in the United States earn substantially less than immigrants of longer U.S. residence. The earnings comparison by degree of permanence suggests two types of Japanese immigrants. A majority who have little incentive to invest in U.S.-specific human capital, and a minority who intend to stay in the United States, and who invest in U.S.-specific human capital.

CONCLUDING REMARKS

The across-group correspondence between measures of permanence and estimates of the return to investment in U.S.-specific human capital suggests that the decision to invest in human capital and the decision to stay permanently in the United States are jointly determined and positively associated. These analyses also alert us to emigration's effect on analyses of immigrant economic assimilation. As all of the Asian developing-country groups have low emigration rates, the potential for emigration to affect their measured earnings growth and human capital investment is limited. Emigration is high, however, for immigrants from Japan, Canada, and English-speaking Western Europe. This suggests caution in following earnings and human-capital-investment measures across censuses for these groups.

The Japanese analysis raises the possibility that the earnings trajectories of the West European groups may be a composite result of two distinctly different groups. A main group, characterized by high entry earnings and low earnings growth and composed of immigrants who do not plan on staying permanently in the United States, and a smaller group, characterized by low entry earnings and high earnings growth, composed of immigrants who intend to stay.

REFERENCES

Ahmed, Bashir and Gregory Robinson, "Estimates of Emigration of the Foreign-Born Population: 1980 1990." *Population Estimates and Projections Technical Working Paper Series*, no. 9, U.S. Bureau of the Census, 1994.

Baker, M. and Benjamin, D. (1997) "The Role of the Family in Immigrants' Labor-Market Activity: An Evaluation of Alternative Explanations," *The American Economic Review* 87(4):705–27.

Beach CM, Worswick C. 1993. "Is There a Double-Negative Effect on the Earnings of Immigrant Women?" *Canadian Public Policy* 19: 36–53.

Chiswick, B.R. (1980) *An Analysis of the Economic Progress and Impact of Immigrants*, Department of Labor monograph, N.T.I.S. No. PB80-200454. Washington, DC: National Technical Information Service.

Chiswick, Barry R. and Miller, Paul W. 1998. "English Language Fluency Among Immigrants in the United States." *Research in Labor Economics*, Vol. 17.

Duleep, H. (1988) *The Economic Status of Americans of Asian Descent: An Exploratory Investigation*, U.S. Commission on Civil Rights, GPO, 1988.

Duleep, H. and Sanders, S. (1993) "The Decision to Work by Married Immigrant Women," *Industrial and Labor Relations Review*, July, pp. 677–690.
Duleep, H. and Regets, M., (1999) "Immigrants and Human Capital Investment," *American Economic Review*, May, pp. 186–191.
Erikson, Charlotte. *Invisible Immigrants: The Adaptation of English and Scottish Immigrants in Nineteenth-Century America.* Coral Gables, Fla.: University of Miami Press. 1972.
Heston, A., R. Summers, "International price and quantity comparisons: potentials and pitfalls." *American Economic Review*, 86 (2) (1996), pp. 20–24.
Long, J. (1980) "The Effect of Americanization on Earnings: Some Evidence for Women," *Journal of Political Economy* 88, 620–629.
MacPherson, David and James Stewart. 1989. "The Labor Force Participation and Earnings Profiles of Married Female Immigrants." *Quarterly Review of Economics and Business*, vol, 29, (Autumn) 57–72.
Piore, Michael J., *Birds of Passage: Migrant Labor and Industrial Societies*, New York: Cambridge University Press, 1979.
Portes A and Bach R. (1985) *Latin Journey: Cuban and Mexican Immigrants in the United States.* Berkeley, Calif.: University of California Press.
Portes, Alejandro and Rafael Mozo. (1985) "The Political Adaptation Process of Cubans and Other Ethnic Minorities in the United States: A Preliminary Analysis," *International Migration Review*, vol. 19, no. 1 Spring.
Reimers, Cordelia W. "Sources of the Family Income Differentials among Hispanics, Blacks, and White Non-Hispanics." *American Journal of Sociology* 89, no. 4 (1984): 889–903.
Schoeni R. 1998. Labor Market Outcomes of Immigrant Women in the United States: 1970 to 1990. *International Migration Review* 32(1): 57–78.

PART III

A Family Perspective

Asian immigrant men from economically developing countries begin their U.S. lives at much lower earnings than their Western European and Canadian counterparts, but experience higher earnings growth. Continuing our focus on immigrants who came to the United States following the 1965 Immigration and Nationality Act, Part III extends the analysis from men to families.[1]

If intergroup variations in the initial earnings of immigrant men stem from the selection of more or less able individuals, and persons of similar ability marry, then the relative economic position of the developing-country Asian groups should worsen when wives' earnings are included. Other theories predict that a family perspective will improve the relative economic status of these groups, either at the beginning of their U.S. journey, or with time in the United States.

The Family Investment Model suggests that financing the husband's investment in host-country skills[2] affects the labor-force and

---

[1] For all foreign born who are 25–64 years old in each census year, the percentage married for the entry cohorts of women are: 77% for the 1965–1970 cohort, 74% for the 1975–1980 cohort, 69% for the 1985–1990 cohort, and 71% for the 1995–2000 cohort. The analogous numbers for men are 78%, 71%, 65%, and 64%. These estimates are based on the 1970, 1980, 1990, 2000 census PUMS files.

[2] The cost of husband's investment in host-country-specific human capital includes both direct and indirect costs in the form of foregone earnings.

human-capital-investment decisions of immigrant women.[3] Specifically, it predicts that: (1) In positive relation to their husbands' investment in host-country skills, women will be more likely to work, work longer hours and more weeks per year than would otherwise be the case. (2) With time in the United States, this greater propensity to work lessens as immigrant husbands acquire host-country-specific human capital. (3) In the initial years following immigration, women will forego human capital investment and pursue jobs that pay more during the time when the husband's investment in host-country human capital is most intense: their initial wages will be higher, than would otherwise be the case, and their wage profile, by foregoing investment, flatter.

A third perspective is that the earnings patterns of immigrant women resemble those of immigrant men. Both will follow an IHCI model wherein (given their initial levels of human capital) earnings growth is inversely related to initial earnings for persons who are permanently attached to the United States.

To further our understanding of immigrant economic assimilation from a family perspective, Chap. 11 measures how family members contribute to family income highlighting the important role women play. Chapter 12 explores possible explanations for the high labor force participation of married women from the Asian developing countries and concludes that conventional explanations cannot explain it. Chapter 13 introduces the Family Investment Model. Following cohorts and individuals over time, Chap. 14 tests the Family Investment Model. Chapter 15 examines the role of unpaid work in family businesses. Chapter 16 steps beyond the confines of the nuclear family to the extended family.

REFERENCES

Baker, M. and Benjamin, D. (1997) "The Role of the Family in Immigrants' Labor-Market Activity: An Evaluation of Alternative Explanations," *The American Economic Review* 87(4): 705–27.

---

[3] Studies that find evidence supporting the Family Investment Model include Long (1980), Duleep and Sanders (1993), Beach and Worswick (1993) and Baker and Benjamin (1997). Other key studies on immigrant women include Chiswick (1980), MacPherson and Stewart (1989), and Schoeni (1998). Reimers (1985) illuminates the prominent role that women's earnings play in family income differences across ethnic groups.

Beach CM, Worswick C. 1993. Is There a Double-Negative Effect on the Earnings of Immigrant Women? *Canadian Public Policy* 19: 36–53.

Chiswick, B.R. (1980) *An Analysis of the Economic Progress and Impact of Immigrants*, Department of Labor monograph, N.T.I.S. No. PB80-200454. Washington, DC: National Technical Information Service.

Duleep, H. and Sanders, S. (1993) "The Decision to Work by Married Immigrant Women," *Industrial and Labor Relations Review*, July, pp. 677–690.

Long, J. (1980) "The Effect of Americanization on Earnings: Some Evidence for Women," *Journal of Political Economy* 88, 620–629.

MacPherson, David and James Stewart. 1989. "The Labor Force Participation and Earnings Profiles of Married Female Immigrants." *Quarterly Review of Economics and Business*, vol, 29, (Autumn) 57–72.

Reimers, Cordelia. 1985. "Cultural Differences in Labor Force Participation among Married Women." *American Economic Review, Papers and Proceedings*, (May): pp. 251–55.

Schoeni R. 1998. Labor Market Outcomes of Immigrant Women in the United States: 1970 to 1990. *International Migration Review* 32(1): 57–78.

CHAPTER 11

Family Income

A family, in the census and in the analyses of Part III, is defined as two or more persons related by birth, marriage, or adoption, who live together as one household. In the analyses that follow, we consider only families in which the household head is between 18 and 65 years old. When we analyze married couple families, both wife and husband are between 18 and 65 and of the same immigrant group.[1] A narrower age range of 25–64 is adopted when we analyze labor force behavior. We begin our exploration with the census-defined family income variable, adjusted for its upper truncation.[2] Then, to elucidate the contributions of each member, we create a measure of family income.

---

[1] Since no information is available on former spouses, the statistics concerning all families (not just married-couple families) are based on samples that include immigrants who were formerly married to U.S.-born individuals. If no spouse is present in the family, the foreign-born status of the family is that of the family's head.

[2] In the census variable, losses exceeding $9990 are recorded only as "greater than $9990" and family incomes exceeding $75,000 are coded as "income of $75,000 or more." To alleviate this shortcoming, we used the Pareto method to estimate the mean family income of families with earnings exceeding $75,000. Given their small number, we assume that families with losses of more than $9990, only lost $9990. The following income values were estimated for family incomes exceeding $75,000: Asian foreign born—$115,877; European and Canadian foreign born—$121,486. The Pareto method is described in Bureau of the Census, *Technical Documentation, 1980 Census*, Appendix J, p. 164. These estimated mean values were assigned to families of each of the groups who reported more than $75,000 of income.

© The Author(s), under exclusive license to Springer Nature Switzerland AG 2020
H. Duleep et al., *Human Capital Investment*,
https://doi.org/10.1007/978-3-030-47083-8_11

Determinants of Family Income

The top half of Table 11.1 restates an earlier finding: Asian immigrant men—other than the Japanese—begin their U.S. economic lives with low earnings relative to European and Canadian immigrant men.[3] Comparing

Table 11.1 Men's earnings and family income by years since migration to the US

Years since migration	Filipino	Chinese	Korean	Indian	Japanese	European and Canadian
Average earnings of immigrant men by years since migration						
1–5	$11,198	$11,156	$12,493	$14,769	$27,112	$17,803
6–10	17,675	16,169	21,726	24,287	21,936	20,533
11 years or more	23,719	20,722	31,647	31,988	24,128	21,751
Relative to European and Canadian immigrant men						
1–5	0.63	0.63	0.70	0.83	1.52	1.00
6–10	0.86	0.79	1.06	1.18	1.07	1.00
11 years or more	1.09	0.95	1.45	1.47	1.11	1.00
Average income of immigrant families by years since migration						
1–5	$21,997	15,596	17,334	20,740	27,347	21,531
6–10	30,471	23,996	27,186	32,198	25,847	24,763
11 years or more	33,039	31,666	37,183	39,918	26,430	28,935
Relative to European and Canadian immigrant families						
1–5	1.02	0.72	0.81	0.96	1.27	1.00
6–10	1.23	0.97	1.10	1.30	1.04	1.00
11 years or more	1.14	1.09	1.29	1.40	0.91	1.00

Source: Authors' estimates. Estimates for the Asian groups are based on the 1980 Census of Population, 5% "A" Public Use Sample; estimates for the European/Canadian group are based on a 1% sample

Notes: The Asian groups are defined by the census race variable, which includes Filipino, Korean, Indian, Chinese, and Japanese as separate groups, and by whether the individual was foreign born (and not born abroad of U.S. parents). The benchmark group includes all Europeans and Canadians. Foreign born are defined as persons born outside of the United States excluding those with U.S. parents. The sample does not exclude students, the self-employed, and persons with zero earnings. The age range for individual men is 25–64 years old whereas the age range for heads of households in the family income comparisons is 18–64; the age range for married women is also 18–64

---

[3] The Asian group definitions used in Chaps. 11, 12, and 15 are somewhat different than the definitions used in the rest of the book. The work in these three chapters builds on prior work by Duleep and Sanders that was focused on differences in economic status by race. The Asian groups are defined by the census race variable, which includes Filipino, Korean, Indian, Chinese, and Japanese as separate groups, and by whether the individual was foreign born (and not born abroad of U.S. parents). Country of origin is used elsewhere. We do not think this difference affects our results for married women but of course welcome further analysis.

family incomes (bottom half of Table 11.1) improves the relative economic standing of the Asian developing-country groups for all year-of-entry cohorts, including recent arrivals.[4] Instead of incomes that are 0.63, 0.63, 0.70, and 0.83 of the European-Canadian benchmark, the relative family incomes of recently arrived Filipinos, Chinese, Koreans, and Indians are 1.02, 0.72, 0.81, and 0.96, respectively. The relative status of Japanese immigrants worsens when families, versus men, are considered.

A similar story emerges with the poverty rate (Table 11.2).[5] Given the low initial earnings of immigrant men in the Asian developing-country groups, one might expect that all groups would have high poverty rates during their initial U.S. years. Yet, except for the Chinese, their rates fall below the rate for recently arrived European and Canadian immigrants.[6]

What explains the improved economic status of the developing-country Asian immigrants when families, as opposed to men, are considered? Family breakups are a major cause of poverty in America.[7] Perhaps the low divorce rates of the developing-country Asian immigrants (Table 11.3) explain the relative improvement.

Families are less likely to be in poverty and have higher average incomes with husband and wife present (Tables 11.4 and 11.5). Yet, the relative income position of the developing-country Asian groups hardly changes when we examine families with intact marriages.[8] With all families considered, the family incomes of developing-country Asian immigrants who have been in the United States five years or less range from 72% to 102%

---

[4] As a point of comparison, the family income in 1979 of U.S.-born non-Hispanic white families in which the head of the household is 18–64 years old was $26,514.

[5] The poverty rate is the percentage of families whose incomes fall below a threshold that considers family size, number of children, and age of the household head. The poverty threshold is based on the Department of Agriculture's 1961 Economy Food Plan and the assumption that one-third of a family's income goes to food. The poverty level is thus three times the current cost of the economy food plan. People below this income level are "poor"; those above it are "not poor." The census variable used to measure the poverty level is the poverty status in 1979 and is defined as the ratio of family income in 1979 to a "poverty threshold." (Bureau of the Census, *Technical Documentation, Census of Population and Housing, 1980*, Appendix K)

[6] As a point of reference, the poverty rate for U.S.-born non-Hispanic white families of the same age range in 1980 was 6.6%.

[7] Mirroring findings for the general population, Simon and Akbari (1996) found family composition to be the principal determinant of welfare use by immigrants, overshadowing other factors such as level of education.

[8] It also appears that marital dissolution is the cause of the unusually high poverty rates of Japanese immigrants who entered the United States before 1975.

Table 11.2 Percentage of immigrant families below the poverty line by year of immigration

Years since migration	Filipino	Chinese	Korean	Indian	Japanese	European and Canadian
1–5 years	8.8%	27.3%	19.4%	14.0%	12.1%	20.4%
6–10 years	3.7	9.6	7.4	4.5	12.3	9.2
11 years or more	3.7	5.1	6.3	2.5	8.4	5.4
Relative to European and Canadian immigrant families						
1–5 years	0.43	1.34	0.95	0.69	0.60	1.00
6–10 years	0.40	1.04	0.81	0.49	1.34	1.00
11 years or more	0.68	0.94	1.16	0.45	1.56	1.00

Source: Authors' estimates. Estimates for the Asian groups are based on the 1980 Census of Population, 5% "A" Public Use Sample; estimates for the European/Canadian group are based on a 1% sample

Notes: The Asian groups are defined by the census race variable, which includes Filipino, Korean, Indian, Chinese, and Japanese as separate groups, and by whether the individual was foreign born (and not born abroad of U.S. parents). The benchmark group includes all Europeans and Canadians. Foreign born are defined as persons born outside of the United States excluding those with U.S. parents. The sample does not exclude students, the self-employed, and persons with zero earnings. The age range for heads of households and married women is 18–64

Table 11.3 Family dissolution among Asian and European-Canadian immigrant groups: percentage of ever-married women who are divorced or separated

	Filipino	Chinese	Korean	Indian	Japanese	European and Canadian
25–64 years old	6.5	4.2	7.2	2.8	9.5	9.6
25–34 years old	6.9	3.5	7.7	2.0	8.1	14.0

Source: Authors' estimates. Estimates for the Asian groups are based on the 1980 Census of Population, 5% "A" Public Use Sample; estimates for the European/Canadian group are based on a 1% sample

Notes: Divorce rates were derived by dividing the number of women who were divorced or separated in 1980 by the number of women in that age category who were ever married. The Asian groups are defined by the census race variable, which includes Filipino, Korean, Indian, Chinese, and Japanese as separate groups, and by whether the individual was foreign born (and not born abroad of U.S. parents). The benchmark group includes all Europeans and Canadians. Foreign born are defined as persons born outside of the United States excluding those with U.S. parents. The sample does not exclude students, the self-employed, and persons with zero earnings. The age range for heads of households and married women is 18–64

Table 11.4 Poverty rates by years since migration, married couples only

Groups	Filipino	Chinese	Korean	Indian	Japanese	European and Canadian
1–5 years	5.0%	26.8%	18.0%	13.1%	9.6%	19.9%
6–10 years	1.5	8.4	4.5	3.5	6.3	8.3
11 or more years	2.5	4.5	3.6	1.9	2.5	4.0
Relative to European and Canadian families						
1–5 years	0.25	1.35	0.91	0.66	0.48	1.00
6–10 years	0.18	1.01	0.55	0.42	0.76	1.00
11 or more years	0.62	1.10	0.90	0.47	0.62	1.00

Source: Authors' estimates. Estimates for the Asian groups are based on the 1980 Census of Population, 5% "A" Public Use Sample; estimates for the European/Canadian group are based on a 1% sample

Notes: The Asian groups are defined by the census race variable, which includes Filipino, Korean, Indian, Chinese, and Japanese as separate groups, and by whether the individual was foreign born (and not born abroad of U.S. parents). The benchmark group includes all Europeans and Canadians. Foreign born are defined as persons born outside of the United States excluding those with U.S. parents. The sample does not exclude students, the self-employed, and persons with zero earnings. The sample is restricted to women who are married to foreign-born men who are in the same racial or country-of-origin census-category/foreign-born group. War brides are not included. The age range for heads of households and married women is 18–64

the corresponding income average of the European and Canadian families. When we consider only married couple families, the range is 70% to 105%.[9]

Family members other than the husband may contribute more to family income in the developing-country Asian groups than in European and Canadian families. To test this hypothesis, we initially divided each family member's earnings by the census family income measure used in the preceding analysis. However, the sum of family members' incomes can easily surpass the census family income truncation point. This variable also includes income other than earned income, such as interest, dividends, and rental income.

We thus created a new measure of family labor income by summing across the earnings of each member.[10] Although each individual's earnings are truncated, our measure has a much higher truncation point than the census family income variable; it eliminates cases in which individual

---

[9] Intergroup differences in family dissolution also contribute to the relatively high poverty rates of some cohorts of Japanese immigrant families. Future analysts will also want to probe the role of refugees in the poverty rates of the Chinese and Europeans.

[10] The individual earnings that we combine are the sum of wage and salary income and farm and non-farm self-employment income, of each family member.

Table 11.5 Average family income by years since migration to the U.S., married couples only

Years since migration	Filipino	Chinese	Korean	Indian	Japanese	European and Canadian
1–5	$23,918	15,933	17,959	21,149	29,284	22,848
6–10	33,548	24,962	29,468	33,335	31,076	25,558
11 years or more	35,369	32,526	41,623	39,600	36,495	30,598
Relative to European and Canadian immigrant families						
1–5	1.05	0.70	0.79	0.93	1.28	1.00
6–10	1.31	0.98	1.15	1.30	1.22	1.00
11 years or more	1.16	1.06	1.36	1.29	1.19	1.00

Source: Authors' estimates. Estimates for the Asian groups are based on the 1980 Census of Population, 5% "A" Public Use Sample; estimates for the European/Canadian group are based on a 1% sample

Notes: The Asian groups are defined by the census race variable, which includes Filipino, Korean, Indian, Chinese, and Japanese as separate groups, and by whether the individual was foreign born (and not born abroad of U.S. parents). The benchmark group includes all Europeans and Canadians. Foreign born are defined as persons born outside of the United States excluding those with U.S. parents. The sample does not exclude students, the self-employed, and persons with zero earnings. The sample is restricted to women who are married to foreign-born men who are in the same racial or country-of-origin census-category/foreign-born group. War brides are not included. Sample sizes for married immigrant women, 18–64: Filipino (3524), Chinese (3855), Korean (1922), Indian (2445), Japanese (821) and European/Canadian (562)

contributions to family income exceed family income itself. To measure the percentage contribution to family income by each member, we divided each person's earnings by total family labor income defined as the earnings of the husband + earnings of the wife + earnings of children + earnings of other relatives.[11] Averaging these individual measures across families by family member type yields an estimate for each immigrant group of the percentage contributions that children, other relatives, and wives make to family income.[12]

On average, children's contributions to family income are quite small. Across immigrant groups, their percentage contribution ranges from 1.2 to 7.2% of family income. Except for the Chinese, it is no greater for the

---

[11] The earnings of any individual, and therefore family income, can be less than zero. The possibility then exists that an individual can have a negative contribution to family income or a contribution greater than all family income or family losses. The measure here of the contribution of family members to family income is truncated so that if an individual earns more than the total family income, that contribution is set to one. Similarly, if the losses of an individual are greater than family income, the losses of the individual are set to negative one.

[12] Only families with both the husband and wife present are considered.

Table 11.6 Contribution of children to family income expressed as a percentage of total family earnings (married couple families only)

	Filipino	Chinese	Korean	Indian	Japanese	European and Canadian
All	4.6	7.2	2.4	3.2	1.2	5.2
By level of husband's earnings						
Less than $5000	11.3	18.1	7.3	9.6	1.8	17.7
5000–9999	6.5	10.3	5.0	4.4	2.9	6.5
10,000–14,999	3.9	7.8	2.6	2.9	2.9	5.4
15,000–19,999	3.2	3.4	1.7	1.7	1.2	4.9
20,000–29,999	3.1	2.1	2.0	0.8	0.5	2.1
30,000–49,999	1.2	1.3	0.9	0.5	0.5	1.4
50,000+	0.9	0.8	0.5	0.4	0.4	0.8

Source: Authors' estimates. Estimates for the Asian groups are based on the 1980 Census of Population, 5% "A" Public Use Sample; estimates for the European/Canadian group are based on a 1% sample. Sample sizes for married immigrant women, 18–64: Filipino (3524), Chinese (3855), Korean (1922), Indian (2445), Japanese (821) and European/Canadian (562)

Notes: The Asian groups are defined by the census race variable, which includes Filipino, Korean, Indian, Chinese, and Japanese as separate groups, and by whether the individual was foreign born (and not born abroad of U.S. parents). The benchmark group includes all Europeans and Canadians. Foreign born are defined as persons born outside of the United States excluding those with U.S. parents. The sample does not exclude students, the self-employed, and persons with zero earnings. Total family earnings here are defined as the sum of all family member earnings including wage and salary income and farm and non-farm self-employment income. The sample is restricted to women who are married to foreign-born men who are in the same racial or country-of-origin census-category/foreign-born group. War brides are not included

developing-country Asian groups than for the Europeans and Canadians (Table 11.6).[13] Children's contributions to family income are important for some immigrant groups when the husband's earnings are low (Table 11.6). There is, however, no evidence of a developing-country Asian effect per se.[14]

---

[13] In examining the role of children in family income, only children who are living at home and who are 18 years of age or younger are considered.

[14] This is true despite the fact that Asian immigrant families are more likely to have children living at home than European and Canadian immigrant families. However, the ratio of working children to all children at home is lower for Asian than for European and Canadian immigrant families. The lower propensity of Asian immigrant children to work than is the case for the European-Canadian group should be more rigorously pursued. The lower propensity of children to work in Asian immigrant families may reflect a greater emphasis on educational activities.

Table 11.7 Other relatives per immigrant family by group: number per 1000 families

	Filipino	Chinese	Korean	Indian	Japanese	European and Canadian
Number of relatives	570	290	230	230	50	110
Ratio of working relatives to all relatives	0.37	0.28	0.22	0.35	0.20	0.27

Source: Authors' estimates. Estimates for the Asian groups are based on the 1980 Census of Population, 5% "A" Public Use Sample; estimates for the European/Canadian group are based on a 1% sample

Notes: Foreign born are defined as persons born outside of the United States excluding those with U.S. parents. The sample does not exclude students, the self-employed, and persons with zero earnings. The Asian groups are defined by the census race variable, which includes Filipino, Korean, Indian, Chinese, and Japanese as separate groups, and by whether the individual was foreign born (and not born abroad of U.S. parents). The benchmark group includes all Europeans and Canadians. Foreign born are defined as persons born outside of the United States excluding those with U.S. parents. The sample does not exclude students, the self-employed, and persons with zero earnings

Siblings, aunts, uncles, parents, and parents-in-law contribute to family income, and the number of adult relatives living with married couple families is at least twice as large among the developing-country Asian groups as it is among Europeans and Canadians (Table 11.7).[15] Relatives' contributions as a percentage of family income are particularly small for Japanese immigrant families, larger for European and Canadian families, and largest for Filipino, Indian, and Chinese immigrant families (Table 11.8). Nevertheless, for all groups, the contribution that live-in relatives make to family income is smaller in importance than that of children. Even disaggregating by husband's income, it never exceeds 7%.

---

[15] The proportion of live-in relatives who work is highest among Filipino and Indian families, for whom more than a third of live-in relatives work. Among Chinese families, the labor force participation of live-in relatives is only slightly higher than for European and Canadian families, while it is lower among Korean and Japanese families (Table 11.7). Across all groups, about a quarter of live-in relatives work. Assuming equal income needs of working and nonworking relatives, this suggests that (in general) the contribution to family income arising from live-in relatives is dominated by the increased burden on family income that their presence engenders.

Table 11.8 Contribution of other relatives to family income expressed as a percentage of total family earnings (married couple families only)

	Filipino	Chinese	Korean	Indian	Japanese	European and Canadian
All	4.15	1.86	1.12	1.62	0.27	1.15
By level of husband's earnings						
Less than $5000	6.9	3.9	1.8	4.0	0	1.8
5000–9999	4.6	2.3	1.0	1.9	0.8	1.2
10,000–14,999	5.1	1.6	1.5	2.2	0.4	1.0
15,000–19,999	3.1	1.8	0.7	1.5	0.4	2.7
20,000–29,999	3.6	0.9	1.3	1.5	0.3	0.5
30,000–49,999	1.2	1.1	0.6	0.5	0	0.1
50,000+	0.4	0.6	0.2	0.2	0	0

Source: Authors' estimates. Estimates for the Asian groups are based on the 1980 Census of Population, 5% "A" Public Use Sample; estimates for the European/Canadian group are based on a 1% sample. Sample sizes for married immigrant women, 18–64: Filipino (3524), Chinese (3855), Korean (1922), Indian (2445), Japanese (821) and European/Canadian (562)

Notes: The Asian groups are defined by the census race variable, which includes Filipino, Korean, Indian, Chinese, and Japanese as separate groups, and by whether the individual was foreign born (and not born abroad of U.S. parents). The benchmark group includes all Europeans and Canadians. Foreign born are defined as persons born outside of the United States excluding those with U.S. parents. The sample does not exclude students, the self-employed, and persons with zero earnings. Total family earnings here are defined as the sum of all family member earnings including wage and salary income and farm and non-farm self-employment income

THE CONTRIBUTIONS OF IMMIGRANT WOMEN

By default, when family earnings are considered, the impressive gain in relative income for the developing-country Asian groups comes from the contributions of immigrant women (Table 11.9). While European and Canadian wives contribute less than 17% of family income and Japanese wives contribute but 9%, their counterparts in the developing-country Asian groups contribute more nearly a quarter; Filipino wives contribute over one-third. This greater role of wives' earnings in the developing-country Asian groups does not simply reflect their husbands' low earnings. Regardless of their husband's earnings level, women in the developing-country Asian groups generally contribute a larger share to family income than their European and Canadian counterparts do.

For immigrants who have been in the United States five years or less, the earnings comparison for women is the reverse of the comparison for men (Table 11.10). Whereas Filipino, Korean, Chinese, and Indian immigrant men earn substantially less than European and Canadian men, their

Table 11.9 Contribution of wives to family income expressed as a percentage of total family earnings (married couple families only)

	Filipino	Chinese	Korean	Indian	Japanese	European and Canadian
All	33.7	23.5	25.1	19.1	8.8	16.6
By level of husband's earnings						
Less than $5000	60.8	39.8	55.5	43.0	31.2	34.4
5000–9999	34.6	26.7	26.6	24.8	22.1	24.0
10,000–14,999	34.1	24.2	24.7	24.5	14.0	22.7
15,000–19,999	32.6	22.1	20.8	19.5	10.8	15.6
20,000–29,999	27.4	18.1	16.8	13.2	4.8	12.5
30,000–49,999	20.3	13.9	13.1	10.6	3.6	5.6
50,000+	14.0	7.7	4.7	10.3	1.2	6.2

Source: Authors' estimates. Estimates for the Asian groups are based on the 1980 Census of Population, 5% "A" Public Use Sample; estimates for the European/Canadian group are based on a 1% sample. Sample sizes for married immigrant women, 18–64: Filipino (3524), Chinese (3855), Korean (1922), Indian (2445), Japanese (821) and European/Canadian (562)

Notes: Total family earnings here are defined as the sum of all family member earnings including wage and salary income and farm and non-farm self-employment income. Foreign born are defined as persons born outside of the United States excluding those with U.S. parents. The sample does not exclude students, the self-employed, and persons with zero earnings. The Asian groups are defined by the census race variable, which includes Filipino, Korean, Indian, Chinese, and Japanese as separate groups, and by whether the individual was foreign born (and not born abroad of U.S. parents). The benchmark group includes all Europeans and Canadians. Foreign born are defined as persons born outside of the United States excluding those with U.S. parents. The sample does not exclude students, the self-employed, and persons with zero earnings

Table 11.10 Earnings of married women and men, 25–64 years old, recent entrants, 1980; year of immigration: 1975–1980

	Filipino	Chinese	Korean	Indian	Japanese	European and Canadian
Married women	$6286	3368	4620	4526	840	2803
Men	$11,198	11,156	12,493	14,769	27,112	17,803

Source: Authors' estimates based on 5% Public Use "A" Sample of the 1980 Census of Population

Notes: Foreign born are defined as persons born outside of the United States excluding those with U.S. parents. The sample does not exclude students, the self-employed, and persons with zero earnings. The Asian groups are defined by the census race variable, which includes Filipino, Korean, Indian, Chinese, and Japanese as separate groups, and by whether the individual was foreign born (and not born abroad of U.S. parents). The benchmark group includes all Europeans and Canadians. The women are married women whose spouses are 25–64 years old, of the same group and nativity. The men include all men, 25–64 years old

Table 11.11 Hours worked and the propensity to work of foreign-born married women, ages 25–64, 1980

	Filipino	Chinese	Korean	Indian	Japanese	European and Canadian
Annual hours	1716.9	1572.10	1660.59	1453.10	1450.22	1434.8
Relative to European-Canadian	1.20	1.10	1.16	1.02	1.01	1.00
Percentage of wives who work	83	65	61	58	26	51
Relative to European-Canadian	1.63	1.27	1.20	1.14	0.51	1.00

Source: Authors' estimates. Estimates for the Asian groups are based on the 1980 Census of Population, 5% "A" Public Use Sample; estimates for the European/Canadian group are based on a 1% sample. Sample sizes for married women, 25–64 years old: Filipino (3342), Chinese (3719), Korean (1815), Indian (2249), Japanese (790) and European/Canadian (531)

Notes: Foreign born are defined as persons born outside of the United States excluding those with U.S. parents. The sample does not exclude students, the self-employed, and persons with zero earnings. The Asian groups are defined by the census race variable, which includes Filipino, Korean, Indian, Chinese, and Japanese as separate groups, and by whether the individual was foreign born (and not born abroad of U.S. parents). The benchmark group includes all Europeans and Canadians

wives earn substantially more than their European and Canadian counterparts. Whereas recently arrived Japanese immigrant men earn substantially more than their European and Canadian men counterparts, their wives earn substantially less.

Among women who work outside the home,[16] married immigrant women in all Asian groups, including the Japanese, work more annual hours than European and Canadian women (Table 11.11). The low relative earnings of Japanese immigrant wives reflect their low propensity to work whereas married women in all the Asian developing-country groups are more likely to work than European and Canadian women (Table 11.11). Filipino women have the highest labor force participation, followed by women from China, Korea, and India.

The high labor force participation of women in the Asian developing-country groups, and the exceptionally low labor force participation of Japanese immigrant women, is surprising given their characteristics

---

[16] Working women are defined as persons who reported positive earnings, positive weeks worked, and positive hours worked in 1979. In other words, women who worked at some point during the year 1979.

Table 11.12 Characteristics of Asian and European/Canadian immigrant married women: 25–64 years old, 1980

	Filipino	Chinese	Korean	Indian	Japanese	European/Canadian
% less than 5 years in US	24	30	48	39	60	15
% less than 10 years in US	60	55	84	81	77	27
% English Proficient[a]	63	29	15	64	15	59
Marital dissolution Rate[b]	6.5	4.2	7.2	2.8	9.5	9.6
Avg # of children at home	2.09	1.77	1.98	1.68	1.43	1.08
% w children under 6	34	28	28	41	38	12
# of children ever born	2.58	2.26	2.21	1.87	1.55	2.36
Avg education	14.42	12.20	13.38	14.33	13.67	10.97
Avg potential Experience[c]	17.90	20.65	17.54	13.83	15.51	26.65

Source: Authors' estimates. Estimates for the Asian groups are based on the 1980 Census of Population, 5% "A" Public Use Sample; estimates for the European/Canadian group are based on a 1% sample

Notes: Includes only immigrant women who are married to men of the same group and nativity. The Asian groups are defined by the census race variable, which includes Filipino, Korean, Indian, Chinese, and Japanese as separate groups, and by whether the individual was foreign born (and not born abroad of U.S. parents). The benchmark group includes all Europeans and Canadians. Estimates based on 5% Public Use "A" Sample of the 1980 Census of Population. Sample sizes for married women, 25–64 years old: Filipino (3342), Chinese (3719), Korean (1815), Indian (2249), Japanese (790), and European/Canadian (531)

[a]A person was classified as English proficient if she reports speaking only English or speaking English very well

[b]Divorce rates were derived by dividing the number of women who were divorced or separated in 1980 by the number of women in that age category who were ever married

[c]Potential experience is measured by a women's age minus her years of schooling minus 6

(Table 11.12). The likelihood of immigrant women working generally increases with time in the United States.[17] Yet, women in the developing-country groups are more likely to be recent entrants than the Europeans and Canadians. One would expect labor force participation to increase with English proficiency; only a small percentage of Chinese and Korean

---

[17]References include Blau (1980), Chiswick (1980), Reimers (1985), Duleep (1988), MacPherson and Stewart (1989).

immigrant women are English proficient. Marital instability increases female labor force participation[18]; women in the developing-country groups have relatively low divorce and separation rates. Pre-school children reduce female labor force participation; the developing-country groups are more likely to have young children at home than their European and Canadian counterparts. The group with the fewest children at home, the Japanese, has the lowest labor force participation. The group with the most children at home, the Filipinos, has the highest labor force participation!

REFERENCES

Blau FD. 1980. "Immigration and Labor Earnings in Early Twentieth Century America," in *Research in Population Economics* (Vol. 2), J Simon, J DaVanzo, eds. Greenwich, Conn.: JAI Press.

Cain, Glen, *Married Women in the Labor Force*, Chicago: University of Chicago Press, 1966.

Chiswick, B.R. (1980) *An Analysis of the Economic Progress and Impact of Immigrants*, Department of Labor monograph, N.T.I.S. No. PB80-200454. Washington, DC: National Technical Information Service.

Duleep, H. (1988) *The Economic Status of Americans of Asian Descent: An Exploratory Investigation*, U.S. Commission on Civil Rights, GPO, 1988.

Johnson, W., & Skinner, J. (1986). "Labor Supply and Marital Separation." *The American Economic Review*, 76(3), 455–469.

MacPherson, David and James Stewart. 1989. "The Labor Force Participation and Earnings Profilesof Married Femlae Immigrants." *Quarterly Review of Economics and Business*, vol, 29, (Autumn) 57–72.

Reimers, Cordelia. 1985. "Cultural Differences in Labor Force Participation among Married Women." *American Economic Review, Papers and Proceedings*, (May): pp. 251–55.

Simon, Julian and Ather Akbari. 1996. "Determinants of Welfare Payment Use by Immigrants and Natives in the United States and Canada," in H. Duleep and P. V. Wunnava (editors), *Immigrants and Immigration Policy: Individual Skills, Family Ties, and Group Identities*, Greenwich, CT: JAI Press, pp. 79–102.

---

[18] See, for instance, Cain (1966) and Johnson and Skinner (1986).

CHAPTER 12

Explaining the High Labor Force Participation of Married Women from Asian Developing Countries

Conceptually, a woman works if the wage she receives (her market wage) exceeds the wage she must receive in order to work (her reservation wage). The reservation wage reflects the monetary and psychic costs of working. Key among these are children. To capture potentially nonlinear effects, we include in our model of female labor force participation categorical variables for whether the family's youngest child is less than 1, 1–5, 6–11, or 12–17 years old. Since older children may care for younger siblings, we include whether children 12 and older are present in homes with children less than 12.

The market wage reflects a woman's skills and the employment conditions of the area in which she lives. We measure general skills with years of schooling (a 2-part spline breaking at 16 years), potential years of work experience (age minus years of schooling minus 6), experience squared, education x experience, and (given the paucity of census work history information) age at first marriage and the number of children ever born. We adjust for employment demand conditions with metropolitan status and region.[1]

---

[1] As health can potentially affect both the reservation wage and market wage, we include a variable measuring the presence of a work-inhibiting disability.

Our basic model is: $P(w) = f(X, Z_1, G)$ where $P(w)$ is the probability that a woman works in paid employment,[2] X and Z_1 represent variables affecting the reservation and market wage, respectively, and G is a set of 0–1 group variables—Japanese, Chinese, Filipino, Korean, and Indian. Married immigrant women from Europe and Canada form the reference group from which the effects of being in the other groups are measured. Within this framework, we examine whether married immigrant women respond to female labor supply variables as do U.S.-born women. Our model's estimated coefficients (Table 12.1) suggest that they do.[3]

More children at home generally lowers labor force participation but varies with children's ages: controlling for the number of children at home, young children decrease while older children increase the propensity to work. The positive estimated effect of children 12 and older in homes with children younger than 12 further suggests that older children help with home activities.[4] The propensity to work increases at a decreasing rate with experience and increases with education. Our splines indicate that additional schooling has a particularly large effect for women with at least 16 years of schooling.[5] Skills transferability permeated our discussion of the earnings of immigrant men.[6] Theoretically, the closer an immigrant's source-country skills are to her U.S. labor market skills, the higher her

---

[2] The dummy variable, whether works, equals 1 if a woman reports positive earnings, positive weeks worked, and positive hours per week worked for the year preceding the census.

[3] Classic studies of the determinants of the labor force participation of married women based on data in which most of the women are native born include Mincer (1962), Cain (1966), and Bowen and Finegan (1969).

[4] Alternatively, the positive coefficient on the presence of both older and younger children may simply signify that women who plan to work space the births of their children. However, we found that this effect persists even in a dynamic analysis (Duleep and Sanders 1994).

[5] In the absence of actual work history data, we added the number of children ever born and age when first married to help control for withdrawals from the labor force. (Controlling for the age structure of children, as we have done, should also help in this regard.) As expected, number of children ever born has a negative effect on the propensity to work. Yet, the coefficient on age at first marriage is small and negative, perhaps reflecting an age-cohort effect.

[6] Previous research suggests that skills transferability is also an important determinant of the labor force behavior of immigrant women. In an analysis of 1970 census data, Chiswick (1980) found that the hourly earnings of immigrant women tend to increase with years since migration. Blau (1980) also found that the earnings of turn-of-the-century immigrant women increased with years in the United States. The earnings growth found in both studies likely reflects a growth in the U.S.-specific capital of immigrant women as they lived in the United States.

12 EXPLAINING THE HIGH LABOR FORCE PARTICIPATION OF MARRIED... 137

Table 12.1 Effects on labor force participation of child status, general skills, demand conditions, and US specific skills: pooled logit model estimated for married immigrant women, ages 25–64

Immigrant groups[a]		Year of immigration[b]	
Filipino	1.7885**	1975–1980	-0.6657**
Chinese	0.6098**	1970–1974	-0.0842**
Korean	0.5544**	1965–1969	0.0396
Indian	0.3502**	1960–1964	-0.1269**
Japanese	-0.9899**	1950–1959	-0.5692**
Children at home		*Ability to speak English*[c]	
Number of children	-0.0793**	Very well	-0.1437**
Youngest child		Well	-0.0158
Under 1	-0.6724**	Not well	0.1362**
Under 6	-0.5378**	Not at all	0.9192**
Under 12	-0.5523*		
Under 17	0.5950**	*Disabled*	-1.5016**
Children under 12 and over 12	0.6438**		
Schooling splines		*Region*[d]	
1–15 years	0.1937**	North central	-0.1559**
16+ years	0.2629**	South	-0.3091**
Potential years of work experience		Part west	-0.2153**
Experience	0.1922**	California	-0.2470**
Experience squared	-0.00259**	Hawaii	-0.3901**
Number of children ever born	-0.0560**		
Age first married	-0.00086	*Location*	
Education x experience	-0.00607**	Outside SMSA	-0.1584**
Intercept	-2.7168**		

Source: Authors' estimates based on 5% Public Use "A" Sample of the 1980 Census of Population. Sample size is 12,446

Notes: Foreign born are defined as persons born outside of the United States excluding those with U.S. parents. Includes only immigrant women who are married to men of the same group and nativity. The Asian groups are defined by the census race variable, which includes Filipino, Korean, Indian, Chinese, and Japanese as separate groups, and by whether the individual was foreign born (and not born abroad of U.S. parents). The benchmark group includes all Europeans and Canadians
[a]The reference category is Europeans and Canadians
[b]The reference category is individuals who immigrated before 1950
[c]The reference category is "speaks only English"
[d]The reference category is East
**Asymptotic t-statistics significant at 0.05 level *Asymptotic t-statistics significant at 0.10 level

market wage, and the higher her propensity to work. Resonating with the skills transferability story, being a recent arrival decreases labor force participation. Contradicting it, the propensity to work increases as English proficiency decreases (Table 12.1).

Separately estimating the model for each immigrant group (Table 12.2) reveals that the negative effect of young children on a wife working is smaller for women in the developing-country groups than for the Japanese and Europeans/Canadians. Having a baby at home has no statistically significant effect for Korean women. Evaluated at sample means, its estimated effect for Filipino women is less than one-fifth the corresponding effect for the Europeans and Canadians.

For all groups, except the Japanese, additional schooling at 16 years and beyond is associated with increases in the propensity to work. For the Japanese, increases in education at less than 16 schooling years decreases the propensity to work.

For European and Canadian immigrant women, not speaking English at all *increases* the probability of working. Rather than implying the irrelevance of English proficiency, this result suggests that levels of English proficiency identify different "types" of immigrants within a given group. That speaking English very well has a large positive effect on the propensity to work for Filipino and Korean women relative to the reference variable—speaking only English—does not negate the importance of English proficiency: immigrants from non-English-speaking countries who report English as their sole language are likely persons who immigrated as children. Our results suggest that women who are immigrants as adults are more likely to work than persons who immigrated as children.[7] Asian Indians show a clear relationship between English proficiency and the propensity to work whereas the relationship for Chinese immigrants is ambivalent.

Controlling for all of these variables, the group effects in our pooled analysis persist. Indeed, for all of the developing country Asian groups, the adjusted group effects exceed the unadjusted effects (Table 12.3).

---

[7] With the exception of Asian Indians, Asian immigrants reporting English as their sole language make up a tiny fraction of the immigrant population and may be persons who immigrated as children. The percentage of women in our sample reporting only English by group is: Chinese—1.6%, Filipinos—1.5%, Indian—8.8%, Japanese—0.9%, and Korean—1.0%.

Table 12.2 Logit coefficients on group-specific estimations for married immigrant women, ages 25–64

	Filipino	Chinese	Korean	Indian	Japanese	European/Canadian
Children at home						
Number of children	-0.1821**	-0.1066**	-0.2080**	-0.0846	-0.2060	-0.0655
Youngest child						
Under 1	-0.2711**	-0.9297**	-0.1336	-0.6329**	-1.1552**	-1.2650*
Under 6	-0.3947	-1.2607**	-0.3142	-0.7209**	-1.7051**	-0.3494
Under 12	0.00907	-0.8171**	0.2109	-0.2369	-1.0910**	-0.8217*
Under 18	0.5377**	0.2333	0.8156**	0.0344	-0.5271	0.6226*
Children under 12 and over 12	0.5100**	0.5549**	0.1694	0.2154	0.5564	0.9100*
Disabled	-1.7584**	-1.3754**	-0.4380**	-0.4356	0.4984	-1.7733**
Schooling splines						
1–15 years	0.0244	-0.0264	0.0204	0.0556	-0.3253**	0.3078**
16+ years	0.1178**	0.1234**	0.1015*	0.2040**	0.1474	0.3943**
Potential years of work experience						
Experience	0.0465	0.0264	0.1642**	0.1755**	-0.2272*	0.2850**
Experience squared	-0.00158**	-0.00097**	-0.00268**	-0.00324**	0.00122	-0.00359**
Number of children ever born	-0.0270	-0.0204	-0.1111*	-0.0494	0.0313	-0.0522
Age first married	-0.00275	-0.00081	-0.0252*	0.0497**	-0.00565	-0.00402
Education X experience	-0.00048	-0.00109	-0.00275	-0.00379*	0.00806	-0.00882**
Region						
North central	-0.1226	-0.3283**	0.3942**	-0.0472	0.0306	-0.1494
South	-0.5482**	-0.0201	0.4033**	-0.1790	0.3673	-0.3005
Part west	0.0453	0.4109**	0.6643**	-0.1169	0.1657	-0.3709
California	0.4020**	0.2794**	0.5361**	0.0142	0.5666**	-0.4912*
Hawaii	0.00293	0.5649**	0.5956*	–	1.8301*	–
Outside SMSA	-0.3405*	-0.3665*	-0.6199*	-0.4040*	-1.2609	0.00327
Year of immigration						
1975–1980	-0.0943	-0.3465*	2.5695**	-0.4928	-1.3552**	-0.8107*
1970–1974	0.5484*	0.3893**	2.6084**	-0.3390	0.2979	-0.1521
1965–1969	0.5485*	0.4848**	2.2025*	-0.7618	0.4649	0.1132
1960–1964	-0.0862	0.3559*	2.1683*	-0.2537	1.2680*	-0.1469
1950–1959	0.1059	0.0924	2.0581*	-0.2204	0.8532	-0.6331**

(*continued*)

Table 12.2 (continued)

	Filipino	Chinese	Korean	Indian	Japanese	European/Canadian
Ability to speak English						
Very well	0.9737**	0.1964	0.9032*	−0.0395	−0.4621	−0.2186
Well	0.8619**	−0.1423	0.5772	−0.4320**	−0.6134	0.1081
Not well	0.7827**	−0.1064	0.4548	−0.7808**	−0.9520	0.3018
Not at all	0.7982	−0.3184	−0.0189	−1.2378**	−1.3982	2.4201**
Intercept	0.4436	1.6803**	−3.3611**	−1.3186	6.5982**	−4.7901**

Source: Authors' estimates for the Asian groups are based on the 1980 Census of Population, 5% "A" Public Use Sample; estimates for the European/Canadian group are based on a 1% sample. Includes only immigrant women who are married to men of the same group and nativity. The Asian groups are defined by the census race variable, which includes Filipino, Korean, Indian, Chinese, and Japanese as separate groups, and by whether the individual was foreign born (and not born abroad of U.S. parents). The benchmark group includes all Europeans and Canadians

Notes: Sample sizes for married women, 25–64 years old: Filipino (3342), Chinese (3719), Korean (1,815), Indian (2249), Japanese (790), and European/Canadian (531)

**Asymptotic t-statistics significant at 0.05 level; *Asymptotic t-statistics significant at 0.10 level

Estimates based on 5% Public Use "A" Sample of the 1980 Census of Population

OTHER EXPLANATIONS

Marriage Before or After Migration

Chiswick theorized that women who migrate on their own accord will be more likely to have readily transferable skills, hence have higher earnings, than otherwise equivalent women who migrate because of their husband's economic opportunities (Chiswick 1980, pp. 189–90).[8] To determine whether a woman married before or after migration, we compare her years since marriage (computed as age minus age at marriage) with the

---

[8] However, his empirical evidence for this hypothesis using 1970 census data was mixed. Non-Hispanic white women whose marriages preceded immigration had hourly earnings that were about 3% lower than other non-Hispanic white immigrant women, other things being equal. For Chinese, Filipino, and Korean women, marriage prior to immigration was not significantly related to earnings. On the other hand, Japanese women who were married prior to immigration were found to earn about 30 percent less than other Japanese immigrant women (Chiswick 1980, p. 212). These results are for women not married to U.S. service men. One of the intriguing results of Chiswick's 1980 study is that immigrant women married to U.S. service men had substantially lower earnings.

Table 12.3 The effect of Asian origin on the labor force participation of married immigrant women (benchmark group is European/Canadian immigrant women)

	Unadjusted Change in Probability	Logit Coefficient[a]	Adjusted Change in Probability
Filipino	0.32	1.7885 (31.16)	0.44
Chinese	0.14	0.6098 (14.38)	0.15
Korean	0.10	0.5544 (9.59)	0.14
Indian	0.07	0.3502 (6.79)	0.09
Japanese	-0.25	-0.9899 (11.07)	-0.24

Source: Authors' estimates based on 5% Public Use "A" Sample of the 1980 Census of Population. The sample size for this pooled regression is 12,446 observations. Includes only immigrant women who are married to men of the same group and nativity. The Asian groups are defined by the census race variable, which includes Filipino, Korean, Indian, Chinese, and Japanese as separate groups, and by whether the individual was foreign born (and not born abroad of U.S. parents). The benchmark group includes all Europeans and Canadians

Notes: The first column of Table 12.3 is the unadjusted difference between the percentage of married women who work in each immigrant group and the corresponding percentage for the benchmark group of European and Canadian women. The estimated logit coefficients are from the model presented in Table 12.1. The third column is the difference in the estimated probability of working between each group and the reference group of Europeans and Canadians, evaluated at the mean probability of working. The effect on labor force participation of changes in an explanatory variable is approximated by $\beta p(1-p)$ where β is the logit coefficient

[a]Asymptotic t-statistics are given in parentheses.

midpoint of her years-since-migration category. Using this criterion for our samples of married immigrant women, roughly 77% of Korean and Indian women, 72% of Japanese women, and from 59 to 61% of Chinese, Filipino, and European/Canadian women married prior to migration. Including this variable in our model (Table 12.4, second column) reveals that women who married prior to migration are *more* likely to work in the United States than those who married after migration; its inclusion slightly increases the coefficients on the economically developing Asian group variables. Group-specific regressions affirm a positive and statistically significant effect of prior-to-migration marriage on the propensity for women to work.[9]

---

[9] Marriage prior to migration was added to the group-specific specifications shown in Table 11.2. In a somewhat different functional form to that presented here, we previously found that the coefficient on whether married prior to migration in an estimation that pooled the immigrant groups was positive, small, and statistically insignificant (Duleep and Sanders 1993). When we estimated a model that interacted each country of origin with marriage

Table 12.4 Selected coefficients from pooled logit model adjusting for whether married prior to migration and adult relatives in the home: 25–64-year-old married immigrant women, 1980

Explanatory variables	Not adjusting for married prior to migration or for adult relatives in the home	Adjusting for married prior to migration	Adjusting for adult relatives in the home
Immigrant groups			
Filipino	1.7885*	1.8290*	1.8114*
Chinese	0.6098*	0.6407*	0.6330*
Korean	0.5544*	0.5841*	0.5744*
Indian	0.3502*	0.3811*	0.3766*
Japanese	-0.9899*	-0.9748*	-0.9723*
Whether married prior to migration		0.4786*	0.4794*
Adult relatives in home			0.0823*

Source: Authors' estimates based on 5% Public Use "A" Sample of the 1980 Census of Population. The sample size for this pooled regression is 12,446 observations. Includes only immigrant women who are married to men of the same group and nativity. The Asian groups are defined by the census race variable, which includes Filipino, Korean, Indian, Chinese, and Japanese as separate groups, and by whether the individual was foreign born (and not born abroad of U.S. parents). The benchmark group includes all Europeans and Canadians.

Notes: All models adjust for child status, general skills, location, and U.S.-specific skills as shown in Table 12.1. The estimated logit coefficients in the first column shown above are from the model presented in Table 12.1

*Asymptotic t-statistics significant at 0.05 level

Relatives in the Home

By helping with child care and other household work, live-in relatives may lower the cost of a woman working outside the home. Hyde (2014, p. 384) comments: "... the heavy reliance by immigrants on family

prior to migration we found that, holding other variables constant, marriage prior to migration was associated with lower labor force participation among Japanese, Chinese, and Filipino women. In contrast, Korean, Indian, and European and Canadian immigrant women who married prior to migration were as likely to or more likely to work as women who migrated before marriage. Whether the difference between these results and the current results reflects some type of error on our part is unknown. Variables that future researchers may want to consider in exploring the effect of marriage prior to migration are the number of years the husband has been in the U.S. and whether the immigrant family intends to stay permanently in the United States.

members for child care evidences a major, although hidden, economic contribution on the part of family members."[10]

Across group variations in the percentage of immigrant homes with live-in relatives match variations in labor force participation: Filipinos are the most likely to have live-in relatives, followed by Chinese, Korean, and Indian households; European and Canadian immigrant households are far less likely to share their home with relatives, and Japanese households even less so. The coefficient on whether there are adult relatives in the home is positive and statistically significant (Table 12.4, third column).[11] Yet, controlling for this variable in our pooled analysis barely affects the Asian group coefficients (Table 12.4). A clue as to why this occurs appears when we interact relatives in the home with each group variable.

Live-in relatives increase the labor force participation of women in all of the Asian immigrant groups but decrease it for European and Canadians perhaps because the relatives themselves require care.[12,13] Evaluated at the sample mean probability, the group coefficients indicate that live-in relatives increase the probability of working 7 to 9 percentage points for Chinese, Filipino, and Korean immigrant women, about 3 percentage points for Indian immigrant women, and a whopping *31 percentage points* for Japanese married immigrant women! Rather than indicating a greater

---

[10] See discussion and references cited in Hyde (2014, pages 380–385). Several researchers have proposed that proximity to relatives contributed to the historically high labor force participation of black women versus white women. In her analysis of late nineteenth-century labor force participation, Goldin (1977, p. 107) writes: "The higher incidence of the extended family structure and a closer community might also be responsible for the smaller response of black women to the presence of young children."

[11] Live-in relatives may also be a financial burden and, via an income effect, increase the likelihood of a woman working. To explore this possibility, we added to our model in another estimation total non-wife family income (including the earnings of other relatives) divided by the number of family members in the home. Our results suggest that the presence of other relatives does not apparently increase an immigrant women's labor force participation by placing a greater burden on family income. If this were the case, we would expect an inverse relationship between the per capita income variable and wife's labor force participation. Instead, the estimated coefficient on this variable is small and positive.

[12] In our analysis, the variables and respective logit coefficients are: Other Relatives, −0.1138; Japanese × Other Relatives, 1.4879; Chinese × Other Relatives, 0.5035; Filipino × Other Relatives, 0.5033; Korean × Other Relatives, 0.4258; Indian × Other Relatives, 0.2367. All of the estimated coefficients are statistically significant.

[13] The above noted diversity of effects of other relatives in the home may be the reason for the variety of estimated effects of relatives in the home on women's labor force participation found in studies of the general population.

effect of live-in relatives on the propensity to work for the Japanese, this large effect suggests that the variable "relatives in the home" identifies different types of immigrants.

Cultural Factors

Cultural factors may underlie the inter-group labor-force-participation variations.[14] Yet, country-of-origin statistics make the intergroup variations we observe to be even more surprising. From 1960 to 1980, women in the Asian developing countries were less likely to work than women in Europe and Canada, whereas women in Japan were more likely to work than women in Europe and Canada (U.S. Bureau of the Census 1992). The fertility rates of women in the Asian developing countries exceeded the U.S. rate; the fertility rate of women in Japan resembled or fell below the U.S. rate (Blau 1992).[15]

Of course, source country statistics may have little to do with "the culture" of those who migrate.[16] To open a window on immigrants' pre-migration past, we limited our 1980 census sample to women who immigrated from 1975 to 1980. For this sample, the census information on whether individuals worked five years ago conveys whether a woman worked in her country of origin before migrating to the United States. Estimating our labor supply model on this subsample (Table 12.5, first column) and adding whether a woman worked full time in 1975 (second column) adjusts for her probable pre-migration work behavior. Consistent with analyses of U.S.-born women, the effect of previous work on current work is large and positive: evaluated at sample means, full-time work prior

---

[14] "Cultural differences may give rise to systematic differences in utility functions that lead to systematic differences in behavior by women in different ethnic or nativity groups who face the same constraints or opportunity set ..." (Cordelia Reimers 1985, p. 251). A cultural assimilation hypothesis has also been discussed and explored with respect to fertility behavior. Refer to Blau (1992) and Kahn (1988).

[15] For instance, Blau (1992) reports a total fertility rate (a proxy for completed fertility) of 3.32 for the U.S., 6.62 for the Philippines, 5.4 for Korea, 6.28 for India, 5.22 for China, and 2.01 for Japan. For the period 1970–1975, total fertility rates were 1.97 for the U.S., 5.49 for the Philippines, 4.43 Korea, 5.64 for India, 3.86 for China, and 2.07 for Japan.

[16] In her analysis of immigrant fertility, Blau (1992) finds that the estimated completed fertility for many immigrant groups falls far below the average for their source country.

Table 12.5 Effect of having worked full-time in country of origin on labor force participation in America: married immigrant women, Ages 25–64, who immigrated in 1975–1980

Explanatory variables	Not Including Previous Work Experience	Including Previous Work Experience
Immigrant groups		
Filipino	2.3433*	2.2072*
Chinese	0.8849*	0.7718*
Korean	1.2623*	1.2579*
Indian	0.8656*	0.9831*
Japanese	-1.1850*	-1.0545*
Whether worked full-time in 1975		1.1483*

Source: Authors' estimates based on 5% Public Use "A" Sample of the 1980 Census of Population. The sample size for married immigrant women, ages 25–64, who immigrated in 1975–1980 is 4225. Includes only immigrant women who are married to men of the same group and nativity. The Asian groups are defined by the census race variable, which includes Filipino, Korean, Indian, Chinese, and Japanese as separate groups, and by whether the individual was foreign born (and not born abroad of U.S. parents). The benchmark group includes all Europeans and Canadians

Notes: Both sets of estimates are based on pooled logit models that adjust for child status, general skills, U.S.-specific skills, as specified in Table 12.1, along with whether married prior to migration and the presence of adult relatives in the home

*Asymptotic t-statistics significant at 0.05 level

to migration raises the probability of working in the United States by 28 percentage points.[17] Nevertheless, including this variable barely budges the group coefficients.

Family Income

Finally, we added to our pooled model three variables that measure the availability and certainty of income other than the wife's—the husband's wage, whether he had been unemployed in 1979, and the family's assets (the sum of their dividends, rents, and interest in 1979). Consistent with

---

[17] References on the strong association between past and present work behavior include Nakamura and Nakamura (1985) and the Spring 1994 issue of the *Journal of Human Resources*.

Table 12.6 The effect of the availability and certainty of income other than the wife's on the labor force participation of married immigrant women

Explanatory variables	No adjustment for husband's wage, unemployment, or family assets	Adjusting for husband's wage, unemployment, and family assets
Immigrant group		
Filipino	1.8114*	1.7363*
Chinese	0.6330*	0.6052*
Korean	0.5774*	0.5286*
Indian	0.3766*	0.3421*
Japanese	-0.9723*	-0.9461*
Husband's wage		-0.00656*
Family assets		-0.00004*
Husband ever unemployed		0.2042*

Source: Estimates based on 5% Public Use "A" Sample of the 1980 Census of Population. The sample size for each pooled regression is 12,446 observations. Includes only immigrant women who are married to men of the same group and nativity. The Asian groups are defined by the census race variable, which includes Filipino, Korean, Indian, Chinese, and Japanese as separate groups, and by whether the individual was foreign born (and not born abroad of U.S. parents). The benchmark group includes all Europeans and Canadians.

Notes: All sets of estimates are based on pooled logit models that adjust for child status, general skills, location, U.S.-specific skills, whether married prior to migration, and the presence of adult relatives in the home. The estimates in the first column are from the same model shown in the third column of Table 12.4

*Asymptotic t-statistics significant at 0.05 level

studies of women's labor force participation, we find that a husband's wage and family assets negatively correlate, and his unemployment experience positively correlates with his wife's propensity to work. Taking these variables into account (Table 12.6) reduces the group coefficients somewhat. Yet, the large group-specific effects persist (Table 12.6).[18]

---

[18] Perhaps variation in permanent income might explain the group differences in women's labor force participation? Yet, from Part I, we know that the incomes of immigrant men across groups converge over time. Permanent income would thus correlate less with the intergroup variation in women's labor force participation than current income. Indeed, when we control for the husband's permanent income, by including husband's education in the labor supply model of married women, the group effects increase.

Permanence

Immigrants who intend to stay in the United States are more likely to invest in U.S.-specific human capital than those who do not consider the U.S. their permanent home. The latter likely include persons who can work here without investing because their jobs do not require U.S.-specific skills (as we hypothesized for the majority of Japanese immigrant men) or because their source-country skills are more generally transferable to the U.S. labor market. These considerations suggest that dividing immigrant families by intent to stay should decrease intergroup variations in the labor force participation of married immigrant women.

The first column of Table 12.7 shows the estimated group coefficients from a pooled model of women's labor force participation that includes all of the explanatory variables we have explored; the second column shows the coefficients from this model estimated on a subset of women whose husbands were U.S. citizens in 1980. Doing this, the effect of Filipino descent decreases and the unexplained effects of Indian and Korean origin are practically eliminated; the formerly large negative effect of Japanese origin, reduced by a factor of five, is no longer statistically significant.

Table 12.7 Selected coefficients from Pooled Probit Model estimated for all married immigrant women and immigrant women married to U.S. citizens (benchmark group is European and Canadian married immigrant women)

Immigrants groups	All immigrants	Permanent immigrants
Filipinos	0.890 (14.95)*	0.671 (7.13)*
Chinese	0.351 (7.58)*	0.318 (4.20)*
Korean	0.367 (5.85)*	-0.016 (0.13)
Indian	0.190 (3.38)*	0.070 (0.62)
Japanese	-470 (4.98)*	-0.089 (0.35)

Source: Authors' estimates based on 5% Public Use "A" Sample of the 1980 Census of Population. The sample size for the pooled regression of all married immigrant women is 12,446 observations. It is 5660 observations for those married to U.S. citizens. Includes only immigrant women who are married to men of the same group and nativity. The Asian groups are defined by the census race variable, which includes Filipino, Korean, Indian, Chinese, and Japanese as separate groups, and by whether the individual was foreign born (and not born abroad of U.S. parents). The benchmark group includes all Europeans and Canadians

*Significant at 0.05 level

Permanence may also clarify the anomalous results for Japanese immigrant women. The very large positive "effect" of live-in relatives on the propensity to work likely stems from the variable "relatives in the home" identifying Japanese families who intend to stay in the United States versus those who do not plan on staying permanently. The odd estimated education effect is likely due to the less educated among Japanese immigrants being more permanent.[19] Similarly, our finding that the propensity to work increases as English proficiency decreases in the pooled model and in the group-specific regression for Europeans/Canadians may be because less English proficient groups are more permanently attached to the United States.

REFERENCES

Blau FD. 1980. "Immigration and Labor Earnings in Early Twentieth Century America," in *Research in Population Economics* (Vol. 2), J Simo, J DaVanzo, eds. Greenwich, Conn.: JAI Press.

Blau FD. 1992. "The Fertility of Immigrant Women: Evidence from High Fertility Source Countries," in *Immigration and the Work Force: Economic consequences for the United States and source areas*, GJ Borjas, RB Freeman, eds. Chicago: University of Chicago Press.

Bowen, W. G., & Finegan, T. A. (1969). *The economics of labor force participation.* Princeton, N.J: Princeton University Press.

Cain, Glen, *Married Women in the Labor Force*, Chicago: University of Chicago Press, 1966.

Chiswick, B.R. (1980) *An Analysis of the Economic Progress and Impact of Immigrants*, Department of Labor monograph, N.T.I.S. No. PB80-200454. Washington, DC: National Technical Information Service.

Duleep, H. and Sanders, S. (1993) "The Decision to Work by Married Immigrant Women," *Industrial and Labor Relations Review*, July, pp. 677–690.

Duleep, H. and Sanders, S. (1994) "Empirical Regularities across Cultures: The Effect of Children on Women's Work," *Journal of Human Resources*, Spring 1994, pp. 328–347.

---

[19] Recall from Chap. 10 that 86% of Japanese managers in the U.S. retain their Japanese citizenship. Japanese immigrant managers who are Japanese citizens earned $33,582 on average in 1980 whereas those who were U.S. citizens earned $22,372.

Goldin, Claudia. 1977. "Female labor force participation: The origin of black and white differences," 1870 and 1880. *Journal of Economic History* 37(1): 87–108.

Hyde, Alan. 2014. "The Law and Economics of Family Unification," *Georgetown Immigration Law Journal*, 28:355–390.

Kahn JR. 1988. "Immigrant Selectivity and Fertility Adaptation in the United States." *Social Forces* 67: 108–27.

Jacob Mincer, (1962) "Labor Force Participation of Married Women: A Study of Labor Supply," NBER Chapters, in: *Aspects of Labor Economics*, pages 63–105, National Bureau of Economic Research, Inc.

Nakamura, A and Nakamura, M. (1985) The *Second Paycheck: A Socioeconomic Analysis of Earnings*, Orlando: Academic Press.

Reimers, Cordelia. (1985) "Cultural Differences in Labor Force Participation among Married Women." *American Economic Review*, (May): pp. 251–55.

CHAPTER 13

Husbands and Wives: Work Decisions in a Family Investment Model?

Thus far, we have estimated a conventional labor supply model modified to include concerns relevant to immigrants:
 W_M=f (Level of Human Capital, U.S.-Specific Skills, Market Conditions)
 W_R=f (the Cost of Working, the Need to Work)
 A woman works if W_M (the market wage) > W_R (the reservation wage) at zero hours of work.
 To this model, we introduce the following concept:

> *Family members can increase future family income by pursuing activities that increase their own skill levels, or by engaging in activities that finance—directly or indirectly—the human capital investment of other family members.*

This line of thought suggests the following reformulation of a woman's decision to work: A woman works if $W_M + E(q) > W_R$ where q is the return in terms of the change in the present net value of family income from the husband's investment in U.S.-specific human capital financed by the wife working: y_I is the husband's earnings stream from human capital investment financed by the wife working, y_{NI} the husband's earnings stream that would exist if the wife did not work, r the market interest rate, and p_t the probability that the family is in the United States in time period t (the probability that they have not emigrated). Then the family investment return to the wife working is $E(q) = \Sigma\ 1/(1+r)^{t-1} \cdot (y_{I,t} - y_{NI,t}) \cdot p_t$.

The price of the wife's non-market activity is not only her market wage, but also the lost investment in the husband's human capital and resulting gain in future family income that would be financed if she worked. In this way, women with identical reservation and market wage characteristics could differ greatly in their propensity to work as a function of the expected return from the husband's human capital investment.

IMMIGRANT WOMEN'S PROPENSITY TO WORK AND THE FAMILY INVESTMENT RETURN

From group-specific regressions for immigrant men, the first row of Table 13.1 gives the approximate proportionate difference between the earnings of men who immigrated during 1975–1980 and the earnings of "otherwise similar" immigrants in the same group who have lived in the United States for 30 or more years. The difference measures the average return to investment in U.S.-specific human capital for each immigrant group; the larger the negative coefficient in the first row of Table 13.1, the

Table 13.1 Percentage impact of years since migration on annual earnings of immigrant men, ages 25–64, by group, 1980

Years since migration	Filipino	Chinese	Korean	Indian	Japanese	European/Canadian
0–5	−0.74*	−0.69*	−0.76*	−0.55*	−0.09	−0.27*
6–10	−0.38*	−0.36*	−0.37	−0.19	−0.17	−0.10
11–15	−0.24*	−0.25*	−0.22	−0.02	−0.07	−0.04
16–20	−0.14*	−0.14*	−0.12	0.04	−0.07	−0.07
21–30	−0.11	−0.12*	−0.19	0.04	−0.12	0.06

Source: Authors' estimates based on a 6% 1980 file created by combining and reweighting the 5% and 1% files of the 1980 Census of Population

Notes: The results presented above indicate the approximate proportionate amount by which the earnings of immigrant men, who immigrated during a specified time period, differ from the earnings of immigrants in the same group who have resided in the U.S. for 30 years or more. The coefficients are derived from group-specific regression estimations in which the dependent variable is the natural logarithm of earnings and the explanatory variables include education, work experience, region, location, marital status, and disability. Estimates based on 5% Public Use "A" Sample of the 1980 Census of Population. The Asian groups are defined by the census race variable, which includes Filipino, Korean, Indian, Chinese, and Japanese as separate groups, and by whether the individual was foreign born (and not born abroad of U.S. parents). The benchmark group includes all Europeans and Canadians

*Significant at 0.05 level

larger the return to investment in U.S.-specific human capital by recently arrived immigrant men.[1]

Comparing these estimates to the group effects on immigrant women's propensity to work discussed in Chap. 12 reveals that the groups with the largest expected growth in immigrant men's earnings (Filipinos, Koreans, and Chinese, followed by Indians) have the highest unexplained labor force participation of married women. The groups with the smallest expected growth in men's earnings (the Japanese and the benchmark group of Europeans and Canadians) have the lowest female participation rates, ceteris paribus.

To test this result in a more precise manner, we use 1990 data for the cohort of married women who immigrated in 1985–1990. For each woman in our sample, we estimate her husband's full potential earnings, $y_{fpi}(h)$, or the earnings he would be expected to have if he possessed U.S.-specific human capital. As a proxy for $y_{fpi}(h)$, we use the earnings of a similarly skilled U.S.-born non-Hispanic white man. Specifically, we calculate average earnings of U.S.-born non-Hispanic white men for 32 earnings groups (four education levels: <8, 9–11, 12–15, 16+; four age ranges: 25–34, 35–44, 45–54, 55–64; and two locations: in SMSA, not in SMSA).[2] We then compute the difference between our estimate of each husband's potential earnings and his earnings for 1989. The gap, $y_{fpi}(h)$-$y_{ci}(h)$, represents the difference between what each husband currently earns, $y_{ci}(h)$, and what he could earn with U.S.-specific human capital—the greater the gap, the greater the return to investment in U.S.-specific skills. Averaging across the individual differences provides an estimate of the average return to human capital investment for married men in each group.

The logit coefficients on the group variables of the women's labor force participation model are in the first column of Table 13.2[3]; their corresponding marginal effects, evaluated at each cohort's pooled mean are in the second column. The third column is the group-specific average return to human capital investment for immigrant husbands.

---

[1] Refer to note 6 in Chapter 10 as to why immigrants who have been in the U.S. 30 or more years are a viable comparison.

[2] Sensitivity analyses revealed that schooling level, age, and residence in a city are the most important factors to predict men's wages.

[3] The group of non-English-Speaking West Europeans has the value of zero since this group is the reference group in the estimation.

Table 13.2 Group effects on the propensity to work of married immigrant women (ages 25–64) who reported migrating to the U.S. in 1985–1990 and each group's average expected return to the husband's investment in U.S.-specific human capital. Immigrant married women from non-English-speaking Western Europe form the reference group (relative ranking in parentheses)

Group	Logit coefficient	Marginal effect	y_p-y_c
Filipino	1.62	0.405 (1)	$27,751 (1)
Chinese	0.790	0.198 (2)	$25,743 (2)
Korean	0.573	0.143 (3)	$22,519 (3)
Indian	0.522	0.130 (4)	$20,186 (4)
Japanese	-1.536	-0.384 (9)	-$17,833 (9)
English-speaking Western Europe	0.230	0.058 (6)	-$11,260 (8)
Other Western Europe	0	0	-$4420 (6)
Eastern Europe	0.331	0.083 (5)	$19,522 (5)
Canadian	0.199	0.050 (7)	-$9332 (7)

Source: Authors' estimates based on a 6% 1980 file created by combining and reweighting the 5% and 1% files of the 1980 Census of Population. Appendix A provides information on sample sizes by entry cohort

Notes: Foreign born are defined as persons born outside of the United States excluding those with U.S. parents. Groups are defined by country of origin. China includes mainland China, Taiwan, and Hong Kong. Korea includes South, North, and unspecified Korea. The sample does not exclude students, the self-employed, and persons with zero earnings. Married women are identified by country of origin; the sample is restricted to women who are married to foreign-born men

The left-hand side of Fig. 13.1 graphs this relationship.[4] The y-axis measures the marginal group effect of married women working; the x-axis is the potential return to human capital investment by their husbands, standardized to a 0 to 100 scale. (We converted our 0–100 scale to a 100–200 scale by adding 100 to each value so that the lowest value, otherwise zero, could be influenced by the probability of staying in the U.S.)

The Family Investment Model we describe suggests that a wife's decision to work is affected by the potential growth in the husband's income, if he invests in U.S.-specific human capital, and the probability of attaining that return, a function of whether the family stays in the United States. To incorporate permanence into our empirical analysis, we weight the husband's average return to investment by our estimates of the probability of

---

[4] We standardized the husband's return to investment to a 0 to 100 scale by adding the lowest value of the average returns shown in Table 13.2 to each value, dividing each value by the highest value, and multiplying by 100.

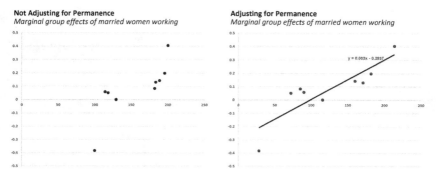

Fig. 13.1 Relative propensity of wife working versus husband's potential return to U.S. investment adjusted for expected permanence: 1985–1990 cohort. (Source: Authors' creation)

staying in the United States for each group. Our measure of permanence is 100 minus the percentage of each cohort who emigrates, as measured by the difference in the number of immigrant men in each age/country cohort counted by the 1980 and 1990 censuses. The right-hand side of Fig. 13.1 weights the potential gains to U.S. investment (measured on the x-axis) by the expected permanence of each group.

Not adjusting for permanence reveals two clusters of points around two almost vertical lines. Incorporating permanence into the analysis transforms the two separate clusters into a single, nearly linear relationship.

CHAPTER 14

Following Cohorts and Individuals Over Time: Work Decisions of Married Immigrant Women

By following cohorts as well as individuals, this chapter tests our findings from previous chapters on the decision to work of married immigrant women. It also extends the prior analyses to their hours worked and wages.

ANALYZING THE DECISION TO WORK BY FOLLOWING COHORTS

The labor supply model presented in Chap. 12 was estimated with cross-sectional data from the 1980 census. This chapter takes that model, minus the English proficiency measures, and uses 1980 and 1990 census data to follow the cohort of married women who immigrated in the years 1975–1980. With 1980 census data, we estimate the model when they are 25 to 54 years old. We then use 1990 census data to estimate the same model when the immigrants of the 1975–1980 cohort are 35–64 years old.

As shown in Table 14.1, women in the Asian developing-country groups are more likely to work during their initial U.S. years than their West European, Canadian, and Japanese counterparts do. This is true even though we have not controlled for English proficiency. Ten years later, the same group differences emerge.[1]

---

[1] Note that the East Europeans, whose husbands have a large initial adjusted earnings gap, resemble the Asian developing-country groups.

Table 14.1 Relationship between time in the United States and the magnitude of the group-specific effects on women's labor force participation; benchmark group is non-English-speaking Western Europe (asymptotic t-statistics in parentheses)

	At entry 1980 (women ages 25–54) who immigrated 1975–1980		10 years later 1990 (women ages 35–64) who immigrated 1975–1980	
	Logit coefficient	Marginal effect	Logit coefficient	Marginal effect
Filipino	1.7420* (13.60)	0.434	2.2567* (13.39)	0.434
Chinese	0.4168* (4.24)	0.104	0.6455* (5.09)	0.124
Korean	0.7585* (7.68)	0.189	0.4504* (3.62)	0.087
Indian	0.5436* (5.46)	0.136	0.4871* (3.66)	0.094
Japanese	−1.6912* (11.30)	−0.396	−0.3964 (1.62)	−0.076
East European	0.3591* (3.76)	0.089	0.7030* (5.26)	0.135
English-speaking West European	−0.1209 (1.02)	−0.030	0.2849 (1.61)	0.055
Canadian	−0.2385* (4.01)	−0.059	−0.0701 (0.36)	0.013

Source: Authors' estimates based on a 6% 1980 file created by combining and reweighting the 5% and 1% files of the 1980 Census of Population and a 6% 1990 file created by combining and reweighting the 5% and 1% files of the 1990 Census of Population. Appendix A provides information on sample sizes by entry cohort at entry and ten years later

Notes: The marginal effects from the logit model are estimated at the mean probability of the sample. Similar results are obtained correcting for sample selection bias. Foreign born are defined as persons born outside of the United States excluding those with U.S. parents. Groups are defined by country of origin. China includes mainland China, Taiwan, and Hong Kong. Korea includes South, North, and unspecified Korea. The sample does not exclude students, the self-employed, and persons with zero earnings. Married women are identified by country of origin. The sample is restricted to women who are married to foreign-born men

*Significant at 0.05 level

Marital Status and Following Cohorts

A problem with following cohorts with census data is that marital status can change: unmarried women in 1980 may marry and be included in the 1990 sample; women who were in the 1980 sample may divorce and not be included in the 1990 sample. Could changes in marital status overturn our comparative results on labor supply?

Divorce rates are lower for the Asian developing-country immigrants than the comparison groups. Johnson and Skinner (1986) suggest that higher divorce risk for women is associated with higher labor supply. For our analysis, in the base year, there is a set of married people with the people who will be divorced in ten years having high labor supply. Ten years later, the divorced women are not part of the married population. This makes the labor force participation of married women ten years later lower (for all groups). If the relationship between the propensity to divorce and labor supply prior to divorce is the same for both groups (and is positive) then the labor force participation of married women will fall the most for the groups with the higher divorce rates. Since Asian women from economically developing countries have low divorce rates, their labor force participation would fall by less than the other groups. This means that the labor force participation of all married groups is lowered by divorce but more so for the groups with more divorce. Thus, the labor force participation gap would grow between developing-country Asian immigrants and the comparison groups—West Europeans, Canadians, and the Japanese.

On the other hand, some of the women who are single in 1980 will be married by 1990. If a high propensity to work is associated with later marriage then as women who marry later join the sample of married women, average labor supply would go up. The marriage rate for Asian immigrants from economically developing countries is higher than for the comparison group. This means that the comparison groups' labor supply would increase relative to Asian immigrants and the gap in labor supply would decrease.

In conclusion, inter-census changes in the composition of marriage could have an effect but the likely pattern is that divorce and marriage partly cancel each other out.

Using Census Data to Follow the Propensity to Work of Individuals Over Time

The Family Investment Model predicts that as their husbands acquire U.S.-specific human capital, the propensity to work of married immigrant women decreases with time in the U.S. relative to immigrant women who are married to men with highly transferable skills. Our results from following the 1975–1980 cohort do not support this prediction. Whereas labor force participation decreases for Korean and Indian women, it increases for the Chinese and shows no relative decrease for the Filipinos (Table 14.1).

To further shed light on this issue, we exploit a feature of the decennial census that allows analysts to follow the same individuals.[2] Specifically, we define a woman as currently working (working in time period t) if, in the 1980 census, she reports working in the 1980 census week.[3] She is defined as working last year (time period t-1) if she reports working in 1979. And, she is defined as working in time period t-5 if she reports working in 1975.[4] Using these definitions, we examine work status changes for U.S. natives and for Asian and European immigrant married women, ages 25–44, who entered the United States in 1975–1980.[5]

Figure 14.1 shows work in the current period conditional on work status in the previous year for married women 25–44, with one child, classified by the age of the child (≤ 1, 2–5, 6–11, 12+) and whether a woman worked the year before. For all groups, women who worked the previous year are very likely to work the current year. The notable exception being mothers with young children, with Asian immigrants being very likely to continue working—despite having young children—and European immigrants and U.S.-born women much less likely to do so. Similar results emerge for women with multiple children.

We also examined work in the current period conditional on work status last year and five years ago. Once again, Asian immigrant women are

---

[2] This analysis is based on Duleep and Sanders (1994). Using census data to create a two-period panel data series for the current and past year work behavior of individuals was initiated by Nakamura and Nakamura (1985). The use of census data to create a two-period panel of work behavior for the last year and five years ago was initiated by Duleep and Sanders (1993). Here we combine these innovations to create a three time period panel.

[3] Note that this is a different dependent variable for whether currently works than we have been using up to this point. Our previous definition of whether a woman works is whether earnings, hours worked, and weeks worked in the preceding year are positive.

[4] The following census questions and responses (U.S. Bureau of the Census) were used to measure (a) employment in the census week (period t), (b) employment during the year prior to the census week (period t-1), and (c) employment five years prior to the census (period t-5):

a. Worked in 1980 census week: *"Did this person work at any time last week?"* We classified a woman as working in period t if the answer is yes.

b. Worked in 1979: *"Last year (1979), did this person work, even for a few days, at a paid job or in a business of farm?"* We classified a woman as having worked in period t-1 if the answer is yes.

c. Worked in 1975: *"In April 1975 (five years ago) was this person—a. on active duty in the Armed Forces? b. Attending college? c. Working at a job or business?—full time; part time?"* We defined a woman as working in period t-5 if it was reported that she worked full or part time in 1975.

[5] Our comparison group is U.S.-born non-Hispanic white women.

Fig. 14.1 Work status in current period for married women 25–44, with one child, classified by age of child and whether worked the year before. (Source: Authors' creation)

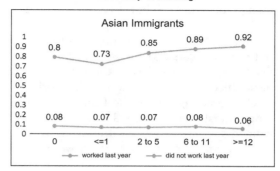

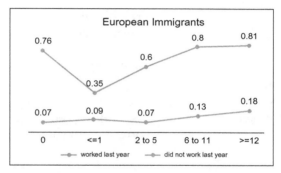

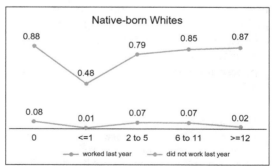

much more likely to continue working with young children than European immigrants and U.S.-born women. More generally, women who worked both last year and five years ago are more likely to work than women who worked in only one of the previous years. Women who worked last year but not five years ago are more likely to work than women who worked five years ago but not last year. The results highlight the importance of persistence in a woman's decision to work.[6]

Contradicting one of the Family Investment Model's predictions, the longitudinal analyses show that the propensity to work of Asian immigrant women does not decrease with time in the United States, either absolutely or relative to European immigrant women and U.S. natives.

Insights on Hours of Work and Wages Following Cohorts

Hours of Work the First Five Years and Ten Years Later

In analyses not shown here, we find that the hours immigrant women work are affected by the same factors and in much the same manner as the decision to work. Hours worked positively correlate with high levels of education, negatively correlate with young children, and positively correlate with older children in the home.

Table 14.2 shows ordinary least squares estimates of the group effects from an hours-of-work model. The first two columns include zero earners in the sample. Echoing our labor force participation results, the developing-country Asian groups work more hours than their West European and Canadian counterparts during the first five U.S. years and ten years later. Japanese immigrant women work much less than women in the Asian

---

[6] To theoretically conceptualize the effect of persistence, consider its effect on a woman's reservation wage. Through several potential routes, the process of working lowers the reservation wage by lowering the cost of working. As one works, preferences are altered as are constraints; a woman who works may become organized to accommodate work, thus the cost of continuing work, or continuing it after a break, is less than the cost of starting work. And, once one has a job, it is easier to continue in that job (or a related one) than beginning afresh. Generalizing from the above findings, the reservation wage would be lower if a woman worked previously and the negative effect on the reservation wage of past work experience would be larger the more recent the work experience, and the more years the woman had worked in the past.

developing-country groups with the differential decreasing with time in the United States.

The greater propensity to work of the Asian developing-country groups does not solely reflect their greater propensity to be in the labor market. The last two columns of Table 14.2 exclude zero earners. Among those who work, the developing-country groups work more hours than their West European and Canadian counterparts, both at entry, and ten years later.[7]

Wages the First Five Years and Ten Years Later

Immigrant men from Asian developing countries start their U.S. careers at much lower earnings than their West European and Canadian counterparts. Yet they experience higher earnings growth. Low-skills transferability, we theorized, underlies their low initial earnings. High expected returns, a low opportunity cost for human capital investment, and the usefulness of source-country human capital for producing new human capital fuel their relatively high earnings growth.

Absent family investment considerations, intergroup differences in the wages of immigrant women might parallel those of immigrant men. Reflecting less transferable skills, the Asian developing country groups should have lower initial wages than West Europeans, with comparable levels of human capital, and higher wage growth. The Family Investment Model predicts the opposite. For groups where the husband has low U.S. skill transferability, married immigrant women would forego human capital investment or delay such investment to take jobs that pay more when their husbands' host-country investment in U.S. human capital is most intense. Asian immigrant women from developing countries should have higher initial wages and flatter wage profiles than their Western European statistical twins.

When we measure the wages of the 1975–1980 cohort of married immigrant women, ages 25–54, during their first five years in the United States (with the 1980 census) and ten years later at ages 35–64 (with the 1990 census), we do not find consistent evidence for the wage trajectory predicted by the Family Investment Model (Table 14.3). Specifically, Korean and Indian immigrant women have high wage growth relative to the West European benchmark.

---

[7] These results persist controlling for selection bias.

Table 14.2 Relationship between time in the United States and the magnitude of the group-specific effects on the hours worked of married immigrant women. Benchmark group is Non-English Speaking Western Europe (t-statistics in parentheses)

	Hours worked (includes zero earners)		Hours worked (excludes zero earners)	
	At entry 1980 (women ages 25–54) who immigrated 1975–1980	10 years later 1990 (women ages 35–64) who immigrated 1975–1980	At entry 1980 (women ages 25–54) who immigrated 1975–1980	10 years later 1990 (women ages 35–64) who immigrated 1975–1980
Filipino	659* (14.42)	787* (13.58)	252* (4.37)	333* (6.00)
Chinese	165* (4.18)	309* (5.74)	81 (1.55)	181* (3.44)
Korean	391* (9.88)	434* (8.02)	224* (4.24)	370* (6.95)
Indian	222* (5.46)	193* (3.35)	106* (1.98)	107** (1.92)
Japanese	-424* (9.03)	-276* (2.39)	-137 (1.28)	-171 (1.42)
East European	50 (1.31)	297* (5.37)	-95** (1.82)	148* (2.74)
English	-166* (3.44)	-52 (0.66)	-261* (3.79)	-209* (2.75)
West European				
Canadian	-133* (2.30)	-111 (1.28)	-162** (1.87)	-12.20 (0.136)

Source: Authors' estimates based on a 6% 1980 file created by combining and reweighting the 5% and 1% files of the 1980 Census of Population and a 6% 1990 file created by combining and reweighting the 5% and 1% files of the 1990 Census of Population. Appendix A provides information on sample sizes by entry cohort at entry and ten years later

Notes: Foreign born are defined as persons born outside of the United States excluding those with U.S. parents. Groups are defined by country of origin. China includes mainland China, Taiwan, and Hong Kong. Korea includes South, North, and unspecified Korea. The sample does not exclude students, the self-employed, and persons with zero earnings. Married women are identified by country of origin. The sample is restricted to women who are married to foreign-born men

*t-statistics significant at 0.05 level; **t-statistics significant at 0.10 level

Concluding Remarks

How the family copes with immigration is important given the low initial earnings of men in several immigrant groups. We find a greater propensity to work for immigrant women from Asian developing countries and, for some groups, high wage growth. We do not find consistent evidence of a family investment strategy in which women pursue low

Table 14.3 Group-specific effects on the wages of married immigrant women during their first five years in the United States, and ten years later, 1980 and 1990 ordinary least squares estimates[a]; Benchmark group is Non-English speaking Western Europe (t-statistics in parentheses)

	Unadjusted Wage Differences	Adjusted Wage Differences	
	At entry 1980 (women ages 25–54) who immigrated 1975–1980	At entry 1980 (women ages 25–54) who immigrated 1975–1980	10 years later 1990 (women ages 35–64 who immigrated 1975–1980)
Filipino	8.47	4.77 (0.82)	0.30 (0.13)
Chinese	-1.56	-2.76 (-0.53)	-0.65 (0.29)
Korean	-0.38	-0.48 (-0.09)	6.04* (2.64)
Indian	-0.17	-3.88 (-0.70)	3.99** (1.70)
Japanese	-0.40	-1.47 (-0.14)	-0.70 (0.14)
East European	-0.74	-0.75 (-0.14)	0.41 (0.18)
English West European	-0.75	-3.99 (-0.54)	-0.26 (-0.08)
Canadian	1.02	-1.54 (-0.17)	2.59 (0.68)

Source: Authors' estimates based on a 6% 1980 file created by combining and reweighting the 5% and 1% files of the 1980 Census of Population and a 6% 1990 file created by combining and reweighting the 5% and 1% files of the 1990 Census of Population. Appendix A provides information on sample sizes by entry cohort at entry and ten years later

Notes: Similar results are obtained correcting for sample selection bias. Foreign born are defined as persons born outside of the United States excluding those with U.S. parents. Groups are defined by country of origin. China includes mainland China, Taiwan, and Hong Kong. Korea includes South, North, and unspecified Korea. The sample does not exclude students, the self-employed, and persons with zero earnings. Married women are identified by country of origin. The sample is restricted to women who are married to foreign-born men

*Significant at 0.05 level; **Significant at 0.10 level

human-capital-investment paths to take jobs that will support their husbands' initial investments in U.S.-specific human capital.[8]

More consistent empirical support rallies around the notion of persistence. Sociological and economic research finds that the experience of working transforms people.[9] Women, who work after migrating—even if

---

[8] Further rebuttals of the Family Investment Model come from Adserà and Ferrer (2014, 2016), Blau et al. (2003), Cohen-Goldner et al. (2009), Cortes (2004), Rashid (2004), and Worswick (1996).

[9] See, for instance Espiritu (1997), Grasmuck and Pessar (1991), Hondagneu-Sotelo (1994), Kibria (1993), and Pessar (1987). Yet, immigrant women in the labor force still have family duties and can carry the double burden of work and household work (Pesquera 1993; Lim 1997).

they did not work in their countries of origin—continue to work. Our research also suggests that a key predictor of work behavior is how permanently attached individuals are to the United States.

We (and others) developed the Family Investment Model to explain large differences across immigrant groups in the work behavior of married immigrant women. Supporting the FIM, groups with the greatest expected return to investment for immigrant men—as measured by an adjusted earnings gap—are the groups where women have the greatest (unexplained) propensity to work in their initial years in the United States. From the work presented here, we cannot conclude that the adjusted earnings gap for immigrant men causally affects the work behavior of immigrant women. Nevertheless, whatever the causality, the strong positive correlation between the size of the adjusted earnings gap for men and the propensity to work of immigrant women is important. For the groups studied in this book, the labor market behavior of immigrant women helps to offset the low initial earnings of the immigrant men who came to the U.S. after the 1965 Immigration and Nationality Act.

References

Alicia Adserà & Ana Ferrer, 2014. "Immigrants and Demography: Marriage, Divorce, and Fertility," Working Papers 1401, University of Waterloo, Department of Economics, revised Jan 2014.

Adserà, A., & Ferrer, A. (2016). "Occupational Skills and Labour Market Progression of Married Immigrant Women in Canada." *Labour economics*, 39, 88–98. https://doi.org/10.1016/j.labeco.2016.02.003

Blau, F., Kahn, L, Moriarty, J. and Souza, A. (2003) "The Role of the Family in Immigrants' Labor-Market Activity: An Evaluation of Alternative Explanations: Comment," *The American Economic Review*, Vol. 93, No. 1 (March) pp. 429–447

Cohen-Goldner, Sarit, Chemi Gotlibovski, and Nava Kahana (2009) "A Reevaluation of the Role of Family in Immigrants' Labor Market Activity: Evidence from a Comparison of Single and Married Immigrants," IZA Discussion Paper No. 4185.

Cortes, K. E. (2004) "Are Refugees Different from Economic Immigrants? Some Empirical Evidence on the Heterogeneity of Immigrant Groups in the United States," *Review of Economics and Statistics*, 2004, 86 (2), 465–480

Duleep, H. and Sanders, S. (1993) "The Decision to Work by Married Immigrant Women," *Industrial and Labor Relations Review*, July, pp. 677–690.

Duleep, H. and Sanders, S. (1994) "Empirical Regularities across Cultures: The Effect of Children on Women's Work," *Journal of Human Resources*, Spring 1994, pp. 328–347.

Espiritu, Yen Le (1997). *Asian American Women and Men*. Thousand Oaks, CA: Sage.

Grasmuck S, Pessar P. 1991. *Two Islands: Dominican International Migration*. Berkeley, Calif.: University of California Press

Hondagneu-Sotelo P. 1994. *Gendered Transitions: Mexican Experiences of Immigration*. Berkeley, Calif.: University of California Press.

Johnson, W.R. and Skinner, J. (1986). "Labor supply and marital separation." *American Economic Review*, 76, 455–469

Kibria N. 1993. *Family Tightrope: The Changing Lives of Vietnamese Americans*. Princeton, N.J.: Princeton University Press.

Lim IS. 1997. "Korean Immigrant Women's Challenge to Gender Inequality at Home: The Interplay of Economic Resources, Gender, and Family." *Gender and Society* 11: 31–51.

Nakamura, A and Nakamura, M. (1985) The *Second Paycheck: A Socioeconomic Analysis of Earnings*, Orlando: Academic Press.

Pesquera, B. (1993). "In the Beginning He Wouldn't Even Lift a Spoon: The Division of Household Labor." in *Building With Our Hands: New Directions In Chicana Studies*. Ed. A. De La Torre and B.M. Pesquera. Berkeley: University Of California Press. Pp. 181–195.

Pessar, P. (1987). "The Dominicans: Women In The Household And The Garment Industry," In *New Immigrants In New York*, Edited By Nancy Foner. New York, NY: Columbia University Press.

Rashid, S. (2004). "Married immigrant women and employment. The role of family investments.". *Umeå Economic Studies,* No 623 http://www.usbe.umu.se/enheter/econ/ues/ues623/'

Worswick C. 1996. "Immigrant Families in the Canadian Labour Market." *Canadian Public Policy* 22: 378–396.

CHAPTER 15

Unpaid Family Labor

Family members may support investment in U.S.-specific human capital as unpaid workers in a family business. By facilitating childcare, the cost of a married woman working in a family business may be less than the cost of her pursuing paid work. Immigrant woman may also choose unpaid work in a family business if their skills transfer poorly to the labor market: these individuals may be productive in a family business. To illuminate this otherwise hidden path of economic assimilation, we use the census work category "unpaid labor in a family business."

Statistics for 1980 suggest that the extent of unpaid labor is small (Table 15.1). When there is unpaid labor, it is likely the wife who is supplying it versus the couple's children or adult live-in relatives,[1] a result consistent with Kim and Hurh's (1986) survey of Korean immigrant businesses.[2] Korean and Chinese immigrant families have the highest levels with 3.6% and 2.1% of immigrant wives reporting that they work without

---

[1] An exception to this statement is the Chinese. Also note that the statistics in Table 15.1 do not reveal to what extent adult relatives are unpaid laborers in homes with adult relatives; the reported statistics also reflect intergroup variations in the extent of live-in adult relatives. In a similar vein, the statistics do not examine unpaid labor from children who are old enough to work. As such, the reported statistics also reflect intergroup variations in the age structure of immigrant families. Future researchers will want to more carefully address these issues.

[2] As discussed below, Kim and Hurh find far higher levels of work participation by wives in the Korean businesses they studied than indicated by the census data.

Table 15.1 Number of unpaid family workers per 1000 families, 1980

	Filipino	Chinese	Korean	Indian	Japanese	European & Canadian
All						
Combined	6.59	39.26	56.75	11.11	18.99	7.53
Wife	4.79	21.24	35.81	8.00	17.72	7.53
Children	0.30	15.33	17.08	1.78	1.27	0.00
Other relatives	1.50	2.69	3.86	1.33	0.00	0.00
Self-employed husbands						
Combined	47.85	147.10	173.23	56.66	108.33	36.15
Wife	38.28	74.97	112.21	43.33	100.00	36.15
Children	0	65.06	49.21	10.00	8.33	0
Other relatives	9.57	7.07	11.81	3.33	0	0

Source: Authors' estimates based on 5% Public Use "A" Sample of the 1980 Census of Population

Notes: Married couple foreign-born families where both husband and wife are ages 25–64. The Asian groups are defined by the census race variable, which includes Filipino, Korean, Indian, Chinese, and Japanese as separate groups, and by whether the individual was foreign born (and not born abroad of U.S. parents). The benchmark group includes all Europeans and Canadians

pay in a family business. Among Asians, these groups also display the highest self-employment rates: 27% of Korean immigrant men and 17% of Chinese immigrant men report self-employment versus 7% and 12% of Filipinos and Indians.[3]

High levels of self-employment need not, however, imply a reliance on family labor. Although 16% of European and Canadian immigrant men report being self-employed, family labor among these immigrants is rare. Limiting the analysis to families with self-employed husbands (second half of Table 15.1): 11% of Korean immigrant wives and 7% of Chinese immigrant wives work without pay in the family business versus less than 4% of Europeans and Canadians.

In contrast to the overall low levels of unpaid family labor reported in the census, Kim and Hurh (1996) find that once a business is in operation, *over half* of the Korean entrepreneurs' wives work an average of 56.6 hours a week in the business with many working 9 or 10 hours a day. The discrepancy between the Kim/Hurh evidence and the census may reflect confusion as to how to report this type of work.

---

[3] Note the relatively high percentage of Japanese women working as unpaid family workers. As a topic for further research, we suspect that this statistic represents the subset of families that are permanently attached to the U.S.

Table 15.2 Percentage of wives working by whether the husband is self-employed, 1980

	Filipino	Chinese	Korean	Indian	Japanese	European and Canadian
All	0.83	0.65	0.61	0.58	0.26	0.51
Husband not self-employed	0.84	0.67	0.64	0.60	0.23	0.53
Husband self-employed	0.67	0.58	0.53	0.43	0.43	0.42

Source: Authors' estimates based on 5% Public Use "A" Sample of the 1980 Census of Population

Notes: Married couple foreign-born families where both husband and wife are ages 25–64. The Asian groups are defined by the census race variable, which includes Filipino, Korean, Indian, Chinese, and Japanese as separate groups, and by whether the individual was foreign born (and not born abroad of U.S. parents). The benchmark group includes all Europeans and Canadians

Table 15.2 shows conventionally defined labor force participation rates of immigrant married women by the husband's self-employment status. For all groups (except the Japanese), women with self-employed husbands are much less likely to work in the paid labor market than women whose husbands are not self-employed. The effect persists in group-specific regressions that include husband's self-employment as an explanatory variable along with all of the explanatory variables discussed in Chap. 12 (Table 15.3).

The negative effect of husband's self-employment on wives' labor force participation may reflect unpaid labor contributed by the wife to the family's business. Indeed, the estimated effects suggest that whether the husband is self-employed is more important to a woman's decision to work than whether he has been recently unemployed and, for some groups, than having young children at home. Underscoring the potential importance of incorporating husband's self-employment status into female labor force participation analyses is the very high self-employment rates of married men in some year-of-entry immigrant cohorts: nearly half of Korean immigrant married men who immigrated between 1965 and 1979 were self-employed in 1990 (Table 15.4).[4]

A more complex picture arises with an analysis of 1990 census data. Having a self-employed husband decreases the probability of working for

---

[4] The statistics in Table 15.4 are tabulated by the wife's year of entry.

Table 15.3 The marginal effects of husband's self-employment (and other variables) on the propensity of married immigrant women to work in the labor market, 1980 and 1990. (Married couple foreign-born families where both husband and wife are ages 25–64)

	Filipino	Chinese	Korean	Indian	Japanese	English-speaking W. Europe	Other West European	Eastern Europe	Canadian
1980 census year									
Effect of youngest child under 1	−0.05	−0.19	−0.03	−0.13	−0.25		−0.32		
Effect of youngest child under 6	−0.07	−0.25	−0.07	−0.15	−0.35		−0.09		
Effect of husband unemployed	0.08	0.07	0.01	0.07	0.03		0.004		
Effect of husband self-employed	−0.16	−0.05	−0.09	−0.10	0.05		−0.11		
1990 census year									
Effect of husband self-employed	0.13	0.03	−0.03	−0.07	−0.12	−0.07	−0.10	0.02	−0.02

Source: Authors' estimates

Notes: The estimations that produced the above results are not exactly comparable. While sharing the same set of explanatory variables, the 1980 estimates are from group-specific probit regressions. The 1990 results are based on a pooled logit model in which husband's self-employment was interacted with each group variable. To facilitate comparing the results of the two analyses, the estimated probit coefficients were scaled by the approximate scaling adjustment proposed by Amemiya (1981, p. 1488). The logit coefficients from the 1990 analysis and the approximated logit coefficients from the 1980 analysis were then converted into marginal effects evaluated at the 1990 mean probability of working. All estimated effects shown in Table 15.3 are statistically significant. Estimates based on 5% Public Use "A" Sample of the 1980 Census of Population and 5% Public Use Sample of the 1990 Census of Population

married immigrant women from Korea, India, Japan, Western Europe, and Canada, but *increases* the probability of working for married immigrant women from China, the Philippines, and Eastern Europe. Why the diversity of results across groups in a given year, and across censuses for the same group?

Ngo (1994) reports that immigrant women in Hong Kong initially work in the paid labor market to help acquire start-up capital to launch family enterprises. In their retrospective survey of Korean immigrant businesses, Kim and Hurh (1996) find that savings from husbands' and wives' earnings, prior to starting a business, constitute the major source of respondents' U.S. capital accumulation: 70% of the spouses were employed before a business opened. Once launched, Korean immigrant wives were heavily engaged in the operation of the family business. Supporting family businesses may include women switching from working in the labor market to help raise funds for family businesses, to working as unpaid labor in the family business.

To test for such a dynamic, we used 1990 census data to estimate the labor supply model for immigrant women that includes (along with all of the explanatory variables discussed in Chap. 12) an explanatory variable that interacts whether the husband is self-employed with his year of immi-

Table 15.4 Percentage of husbands who are self-employed by year of entry, 1990 (Married couple foreign-born families where both husband and wife are ages 25–64)

Year of entry	Filipino	Chinese	Korean	Indian	Japanese	English-speaking West European	Other West European	Eastern Europe	Canadian
All	7.81	17.06	36.67	15.48	9.63	13.51	17.93	14.04	18.63
1985–1990	3.50	8.45	20.60	8.43	2.54	9.63	9.96	7.07	11.88
1980–1984	4.62	19.98	40.28	12.34	18.45	13.40	16.67	14.73	17.07
1975–1979	6.18	21.97	45.82	20.72	28.41	17.81	15.27	21.10	30.81
1970–1974	10.86	21.64	48.13	20.90	40.96	14.65	18.21	17.55	26.04
1965–1969	14.44	21.46	49.02	21.31	32.35	13.97	18.49	14.66	21.60

Source: Authors' estimates based on 5% Public Use Sample of the 1990 Census of Population

gration. The estimated interactive effects suggest an initial positive effect of husband's self-employment on a wife's propensity to work in the labor force, followed by increasingly larger negative effects. The estimated effects range from a two-percentage-point increase in the probability of working, for wives of recently arrived self-employed husbands, to a nine-percentage-point decrease, for wives of self-employed husbands who, in 1990, had been in the United States for 15 to 20 years.[5] The effect of husband's self-employment on a woman's propensity to work varies not only by immigrant group but also with time in the United States. Changes in the years-since-migration distribution for a given group could explain why the estimated effects of husband's self-employment differ for the same group between the 1980 and 1990 censuses.

As an issue for further research, one would theoretically expect that the nature of work and the timing of any changes would depend on the individual skills of immigrant women (and their husbands). The higher the wife's own skill transferability to the U.S. labor market, the more likely she will support a family business through paid versus unpaid labor. This may explain some of the variations among the self-employed in the prevalence of unpaid labor across immigrant groups (second half of Table 15.1). There are relatively high levels of unpaid family labor in groups with low English language proficiency (the Koreans, the Chinese, and the Japanese) and relatively low levels in groups with high levels of English proficiency (Filipinos, Indians, and Europeans/Canadians).

Another area warranting further exploration is to what extent other family members support investment in U.S.-specific human capital by providing child care and household labor while the wife works for pay. It is unlikely that forms of unpaid family work such as child care or domestic work are recorded as "unpaid family work"; this leaves the possibility that other members of the family contribute to U.S.-specific human capital by taking on these tasks. This topic is also relevant to the issue of whether live-in relatives are a net burden on family income as discussed in Chap. 11.

---

[5] There are, from this initial exploration, a number of empirical paths worth pursuing. Our estimates are based on a cross-section. Future researchers will want to explore the effects of husband's self-employment on the wife's propensity to work following cohorts, and if possible, individual families over time. Moreover, we have assumed in our analysis that the time-path effects are the same across immigrant groups.

Concluding Remarks

Wives, children, and other relatives may contribute to family-owned businesses as unpaid workers. As they are unpaid, their contribution to the family's economic assimilation is not reflected in the family earnings studied in Chap. 11. Although not generally important for immigrants, among the self-employed, unpaid labor is important for some immigrant cohorts. Moreover, whether the husband is self-employed affects the wife's decision to work in the labor market. Its effect often exceeds the effects of other husband-related characteristics typically found in female labor force participation models.[6]

References

Amemiya, T. "Qualitative Response Models: A Survey," *Journal of Economic Literature*, vol. 19, December 1981.

Kim KC and Hurh WM. 1996. "Ethnic Resources Utilization of Korean Immigrant Entrepreneurs in the Chicago Minority Area," in *Immigrants and Immigration Policy: Individual Skills, Family Ties, and Group Identities*, HO Duleep, PV Wunnava, eds. Greenwich, Conn.: JAI Press.

Ngo, Hang-yue (1994) "The economic role of immigrant wives in Hong Kong." *International Migration*, Vol. 32, No. 3, 403–23.

---

[6] As with immigrants, few U.S.-born individuals report unpaid labor to the census. Analysis of 1980 census data reveals that 1.3% of U.S.-born, non-Hispanic white married women work as unpaid labor; among those with self-employed husbands, 6% report supplying unpaid labor to a family business. Yet, husband's self-employment has a large negative effect on a native women's propensity to work, perhaps reflecting greater household duties or the wife's actual participation in the operation of a family business. Among U.S.-born non-Hispanic white married women who were not married to self-employed husbands, 60% reported working in 1980, compared with 52% of those with self-employed husbands. A husband's self-employment or plans for a business may also increase a native women's propensity to pursue paid work to financially support an entrepreneurial endeavor. These issues merit greater attention in models of female labor force participation.

CHAPTER 16

Beyond the Immediate Family

To the extent that members of the extended family or community facilitate investment in human capital, we would expect immigrants with extended family to have earnings profiles characterized by lower initial earnings and higher earnings growth than immigrants lacking U.S. extended family.[1] To test this hypothesis, we would like to estimate:

$$y_i = \alpha + \gamma F + X'\beta_1 + (\beta_2 + \Theta F)\text{YSM} + \varepsilon_i$$

where y_i denotes the earnings of immigrant i[2]; X is a vector of the variables education, experience, and region of residence, and β_1 the corresponding

---

[1] Please refer to discussion and references cited in Section B "Family Members Aid in the Growth of Family Businesses and Receive Help in Starting Businesses of Their Own" of Hyde (2014, pp. 385–387). Case studies provide evidence of immigrant communities contributing to a tradeoff between the initial earnings and earnings growth of immigrants (e.g. Gallo and Bailey (1996), Bailey (1987), Waldinger (1986), Portes and Bach (1985), Light (1972)). These studies document an immigrant sector in various industries characterized by mutually beneficial arrangements between recent immigrants and longer term immigrants in which recent immigrants working as unskilled laborers at low wages (or even no wages) in immigrant run businesses are provided training and other forms of support eventually leading to more skilled positions or self-employment.

[2] We use the natural logarithm of annual earnings, which in effect means that we exclude zero earners. Please refer to Chap. 5: Methodological Implications of a Human Capital Investment Perspective for a discussion of the problems associated with excluding zero earners.

coefficients; YSM measures years since migration; and F measures whether immigrant i has extended family members in the United States. If extended family facilitates investment then γ should be negative—lower initial earnings due to greater investment—and Θ should be positive—higher earnings growth due to greater investment for immigrants with extended family.

Census data do not identify family relatives beyond the walls of an individual's home. To proxy the presence of extended family members, we use annual records from the former Immigration and Naturalization Service to compute the percentage of each country-of-origin/year-of-entry cohort that gained U.S. admission via the sibling preference category. We then match this information to 1990 census data on individuals by year of entry and country of origin. The higher the percentage of each cohort that gained entry via the sibling category, the greater the likelihood that an immigrant has relatives in the United States, or is a member of an immigrant community nurtured by kinship admissions.[3]

Using sibling admissions as a proxy for the presence of extended family, our estimating equation becomes:

$$y_i = \alpha + \gamma_1 \text{PerSib}_{jk} + X' \beta_1 + \left(\beta_2 + \Theta \text{PerSib}_{jk}\right) \text{YSM} + \varepsilon_i$$

where PerSib_{jk} is the percentage of immigrants in group j and cohort k who were admitted through the sibling-preference category. YSM denotes years since migration and the vector X includes age, age squared, and seven education categories.[4] The estimated coefficients from this model for immigrant men are shown in the first column of Table 16.1.

---

[3] In explaining individual immigrant behavior, the variable, percentage of cohort$_{jk}$ that entered via the siblings' admission criteria is used as a proxy for whether individual immigrants had extended family in the United States. As such, it is subject to measurement error. To the extent that this measurement error is random, it leads to an understatement of the true effects of having siblings in the country. The variable may be more accurately thought of as the extent to which the community an individual immigrant is in is kinship-based. Another problem with analyses in which group-level information is analyzed at the individual level is that the t-statistics of estimated coefficients may be upward biased. Refer to Moulton (1986) for a discussion of this problem and a possible estimation solution. Given the magnitude of the t-statistics on our variables of interest, we are not concerned about these variables becoming statistically insignificant taking into account the concerns raised by Moulton.

[4] In the 1990 data, we used age instead of the usual experience proxy of age minus years of schooling minus 6 because the 1990 census schooling information is given in categories (e.g. bachelor's degree) instead of years of schooling as is the case in earlier censuses.

Table 16.1 Regression of log (earnings) for immigrant men, 25–64 years old, controlling for percentage of country-of-origin/year-of-immigration cohort admitted via the sibling preference (t-test statistics are in parentheses)

Explanatory variables	No control for percent occupation admissions	Controls for percent occupational skills admissions (versus family admissions in general)		
	All immigrants	All immigrants	Asian immigrants	European immigrants
Years since migration	0.0254* (27.71)	0.0467* (45.79)	0.0656* (26.42)	0.0622* (19.58)
Percent admitted via the sibling preference	-1.0506* (-19.25)	-0.9477* (-17.81)	-0.9425* (-11.51)	0.7037* (3.95)
Years since migration x percent sibling preference	0.1322* (25.57)	0.1065* (19.97)	0.0800* (9.45)	-0.0637* (-4.34)
Percent admitted on the basis of occupation		3.5189* (58.23)	4.1912* (38.51)	2.9178* (22.83)
Years since migration x percent occupation		-0.1333* (-17.80)	-0.2588* (-19.19)	-0.1548* (-9.94)
Adjusted R^2	0.21	0.25	0.22	0.17
Sample size	112,253	112,253	42,723	18,724

Source: Authors' estimates based on a 6% microdata sample created by combining and reweighting the 1990 Census of Population Public Use 5% and 1% Public Use samples

Notes: The regressions also included the explanatory variables age, age squared, and eight categorical variables (including the reference category) measuring the education level of immigrants

*Statistically significant at 0.05 level assuming independent error terms

The estimated coefficient on Percent Admitted via the Sibling Preference suggests that a 10 percentage point increase in sibling admissions lowers the initial earnings of immigrant men by 10.5%. According to the estimated coefficient on the interactive term (PerSib x YSM), the same 10 percentage point change increases the growth in annual earnings by 1.3% per year. The lower initial earnings and higher earnings growth in cohorts with high sibling representation suggest that extended families facilitate investment in U.S.-specific human capital.

Yet, cohorts with high sibling admissions also have high family admissions: the coefficients on sibling admissions may be capturing the earnings effect of individual skill transferability associated with family admission (Chap. 8). To control for this possibility, we added to the above estimation the percentage of occupational-skills admissions in each source-country/year-of-entry cohort, alone and interacted with years since migration (second column of Table 16.1). Though decreasing the magnitude of the sibling-admissions coefficients, strong and significant effects on initial earnings and earnings growth persist.

When we estimate this model separately for Asian and European immigrant men (third and fourth columns of Table 16.1), employment admissions positively affect initial earnings and negatively affect earnings growth for both groups. In other words, family admissions negatively affect initial earnings and positively affect earnings growth. For Asian immigrants, sibling admissions negatively affect initial earnings and positively affect earnings growth, consistent with the thesis that extended family, or being part of a group nurtured by extended family admissions, aids the human capital investment of individual immigrants. For European immigrants, the estimated coefficients suggest the opposite. As in the unpaid family labor results of Chap. 15, and the effect of live-in relatives on the labor force participation of women in Chap. 12, we find, once again, that the role of families in immigrant economic assimilation differs between European immigrants and the developing-country Asian groups.[5]

THE EXTENDED FAMILY AND IMMIGRANT SELF EMPLOYMENT

Case study evidence suggests that extended families play a major role in the high self-employment rates of certain immigrant groups. Madhulika Khandewal (1996) cites examples within the U.S.-Indian immigrant community.[6]

---

[5] An alternative or additional possibility for the differential responses is that sibling admissions may be picking up the effect on earnings of lower education among European immigrants. Sibling admissions are positively associated with educational attainment for Asian immigrants, whereas the reverse is true for European immigrants (Duleep and Regets, 1996). Although education is controlled for in our estimation, it is not interacted with initial earnings and earnings growth as are our measures of extended family.

[6] Although the overall self-employment rate of Indian immigrants is not particularly high, it is high for certain Indian ethnic groups: we estimate that in 1990, nearly a fifth of Asian Indian immigrants in the United States who speak Gujarati were self-employed.

Perhaps the most notable example [of extended family entrepreneurship] is in the motel industry where immigrants from Gujarat, India, and within them a particular subcaste with the last name "Patel," have made their presence felt. A chain of motels owned by these extended families may be found all over the country, even in its remotest parts .... New positions in their growing motel businesses may be filled from the pool of the extended family, comprised of brothers and brother-in-laws, sisters and sister-in-laws, cousins and uncles. Through such occupational networks, the family provides not only employment opportunities but also social and economic support in times of need.

Closeness of family and kin is apparent in yet another Indian concentration in New York City ethnic businesses—the jewelry trade. In a quiet but systematic process of over two decades, immigrants from families engaged in the jewel business traditionally in the western Indian state of Gujarat (and actually originating from a small town called Palanpur), have become the second largest ethnic group in the diamond district in New York City. ... Applying their business acumen and drawing on family and kin resources, they have captured the American part of the worldwide trade in small cut diamonds.

Kim and Hurh (1996) document the important role relatives play in the entrepreneurial ventures of Koreans. Beyond the family, the existence of close-knit communities—the development of which would be aided by kinship admissions—has been hypothesized to promote immigrant entrepreneurial activities (e.g., Bonacich and Modell (1980), Kim, Hurh, and Fernandez (1989), Light (1972), and Waldinger (1989)). Extended immigrant families and close-knit immigrant communities may also facilitate immigrant business formation by providing a supply of trusted employees and informal enforcement mechanisms that decrease variance in employee performance (Duleep and Wunnava 1996; Jiobu 1996).

To statistically assess the potential effect extended family may have on immigrant self-employment, we estimated the correlations between the percentage of each country-of-origin/year-of-entry cohort that was self-employed and, respectively, the percentage admitted on the basis of occupational skills and the percentage of sibling admissions (Table 16.2). There is no statistically significant correlation, for either Asian or European immigrant men, between whether an immigrant is self-employed and the percentage of his year-of-immigration/country-of-origin cohort that was admitted via occupational skills.[7] However, for both groups, there is a positive and highly statistically significant relationship between the propensity of individual immigrants to be self-employed and the percentage of their cohort that gained admission as siblings.

To probe this relationship further in a multivariate context, we estimated for Asian and European immigrant men the following model:

$$P(\text{SelfEmp})_i = \alpha + X'\beta_1 + \beta_2 \text{YSM} + \gamma_1 \text{PerSib}_{jk} + \gamma_2 \text{PerOcc}_{jk} + \varepsilon_i$$

$P(\text{SelfEmp})_i$ is the probability that immigrant i is self-employed, PerSib_{jk} is the percentage of immigrants in group j and cohort k who were admitted through the sibling admission category, PerOcc_{jk} is the percentage of immigrants in group j and cohort k who were admitted through the occupational-skills-preference categories, YSM is years since migration, and the vector X includes age, age squared, and seven education categories (0–8 years of schooling serves as the reference category). The estimated coefficients from the logit model are shown in Table 16.3.[8]

Although our focus is sibling-admission effects on self-employment, the coefficients on the human capital variables are of interest. Had we simply included education as a linear variable, we would have estimated a

Table 16.2 Admission criteria and the propensity to be self-employed: Correlation coefficients (p values in parentheses)

	Asia	Europe
Admission criteria		
Percent admitted on the basis of occupational skills	0.0050 (0.9132)	0.0006 (0.9337)
Percent admitted as siblings of U.S. citizens	0.1335* (0.0001)	0.0453* (0.0001)

Source: Authors' estimates based on a 6% microdata sample created by combining and reweighting the 1990 Census of Population Public Use 5% and 1% Public Use samples

Notes: *A p value of 0.0001 indicates that had the true correlation between the specific admission criteria and the propensity to be self-employed been zero, there would have only been a 0.0001 probability of obtaining correlation coefficients of this magnitude

[7] The p values in parentheses in Table 16.2 indicate statistical significance. A p value of 0.0001 on the siblings correlation coefficient of 0.1335 for Asians indicates that the probability of the true correlation being zero is less than .0001; a p value of 0.9132, as shown on the occupational-skills correlation coefficient of 0.005 for Asians, indicates that there is at least a 90% probability of measuring a correlation of that magnitude given the absence of a correlation between employment-based admissions and self-employment in the population.

[8] The logit model was estimated by maximum likelihood estimation.

Table 16.3 Logistic model of immigrant self-employment (asymptotic t-test statistics are in parentheses)

	Asia	Europe
Explanatory variables		
Intercept	-7.8395 (25.17)*	-4.4680 (10.40)*
Schooling: 9–11 years of schooling	0.3770 (4.86)*	0.2359 (2.26)*
12 years	0.5007 (8.13)*	0.2781 (2.98)*
Some college	0.3646 (6.02)*	0.3732 (3.97)*
Bachelor's degree	0.2384 (3.98)*	0.0865 (0.85)
Master's degree	-0.0547 (0.78)	-0.0674 (0.63)
Professional degree	0.9565 (12.39)*	0.6987 (5.68)*
PhD	-0.6679 (6.09)*	-0.8865 (5.63)*
Age	0.2017 (13.91)*	0.0737 (3.58)*
Age^2	-0.00208 (12.84)*	-0.0008 (3.43)*
Years since migration	0.0757 (20.74)*	0.1154 (18.73)*
Percent admitted on the basis of occupational skills	-0.4282 (1.97)*	0.5854 (2.94)*
Percent admitted as siblings of U.S. citizens	2.7850 (24.47)*	-0.9151 (4.87)*
Sample size	42,723	18,724

Source: Authors' estimates based on a 6% microdata sample created by combining and reweighting the 1990 Census of Population Public Use 5% and 1% Public Use samples

Note: *Statistically significant at 0.05 level assuming independent error terms

positive coefficient indicating that the propensity to be self-employed steadily increases with years of schooling.[9] Our more detailed exploration reveals a nonlinear relationship.[10] For Asian immigrants, the propensity to be self-employed increases until 12 years of schooling and then, with the exception of professional degrees, decreases: controlling for age, those with less than nine years of schooling are as likely to be entrepreneurs as those with a Master's degree; Asian immigrants with a PhD are much less likely to be self-employed than are all other education categories, including the reference group of immigrants with only 0–8 years of schooling.

---

[9] In the 1990 census, respondents were not asked about their years of education, but the grade or degree level they had completed. While this is problematic for the traditional estimation of rates of return to a year of education in human capital earnings functions, it is a useful breakdown in analyses such as this.

[10] The nonlinear relationship between education and self-employment may occur because many jobs in education require a MA or PhD, and educators are not self-employed.

The European pattern is similar except that the peak occurs for immigrants with some college. For both groups, professional degrees (medical doctors and lawyers) are the exception to the low self-employment among the highly educated.

The estimated coefficients on years since migration and age accord with human capital theory. We would expect a positive association between years since migration and the likelihood of being self-employed since newly arrived immigrants likely lack both the capital and information to start a business.[11] Controlling for years since migration, the coefficient on age is measuring the effect on self-employment of age at immigration. Not surprisingly, given the capital requirements of starting a business, the propensity to be self-employed increases with age. Consistent with the estimated negative coefficient on the age-squared variable, we would expect a decrease in the rate at which the propensity to be self-employed increases with age of immigration: the time to reap the benefits of an entrepreneurial venture decreases the older the age of arrival.

Once we control for education, age, and years since migration we find, for Europeans, that sibling admissions negatively affect and employment admissions positively affect the propensity to be self-employed. The opposite holds for Asian immigrant men: employment admissions negatively affect the propensity to be self-employed and the effect of sibling admissions is positive and large. Indeed, for Asian immigrant men, the effect of sibling admissions on the propensity to be self-employed exceeds that of any other variable in the analysis.[12]

---

[11] The work of Kim and Hurh (1996) suggests that immigrant networks may result in a fairly short start-up time. We should also note that had we included a squared years-since-migration term in our estimating equation, we would expect a negative sign for the same reasons as those discussed below with respect to the age-squared explanatory variable.

[12] This finding concurs with that of Fawcett, Carino, Park and Gardner from their longitudinal survey of Korean and Filipino immigrants. In summarizing their results they write: "Most striking ... is the very high level of Economic Contribution scores for Korean family preference immigrants in comparison to occupational preference immigrants. The component scores ... suggest that this difference is related to the greater propensity of Korean immigrants to go into business for themselves" (1990, pp. 13–14).

References

Bailey, Thomas R. (1987) *Immigrant and Native Workers: Contrasts and Competition*, Conservation of Human Resources, Boulder and London: Westview Press.

Bonacich E, Modell J. 1980. *The Economic Basis of Ethnic Solidarity: Small Business in the Japanese-American Community*. Berkeley, Calif: University of California Press.

Duleep, H. and Regets, M. (1996) "Earnings Convergence: Does it Matter Where Immigrants Come From or Why?," *Canadian Journal of Economics*, vol. 29, April.

Duleep, H. and Wunnava, P.V. (1996) Immigrants *and Immigration Policy: Individual Skills, Family Ties, and Group Identities*, Greenwich, Conn.: JAI Press.

Fawcett JT, Carino BV, Park IH, Gardner RW. 1990. "Selectivity and Diversity: The Effects of U.S. Immigration Policy on Immigrant Characteristics." Paper presented at the annual meeting of the Population Association of America.

Gallo C, Bailey TR. (1996) "Social Networks and Skills-Based Immigration Policy," in *Immigrants and Immigration Policy: Individual Skills, Family Ties, and Group Identities*, HO Duleep, PV Wunnava, eds. Greenwich, Conn.: JAI Press.

Hyde, Alan. 2014. "The Law and Economics of Family Unification," *Georgetown Immigration Law Journal*, 28:355–390.

Jiobu RM. 1996. "Explaining the Ethnic Effect," in *Immigrants and Immigration Policy: Individual Skills, Family Ties, and Group Identities*, HO Duleep, PV Wunnava, eds. Greenwich, Conn.: JAI Press.

Khandelwal M. 1996. "Indian Networks in the United States: Class and Transnational Identities," in *Immigrants and Immigration Policy: Individual Skills, Family Ties, and Group Identities*, HO Duleep, PV Wunnava, eds. Greenwich, Conn.: JAI Press.

Kim KC and Hurh WM. 1996. "Ethnic Resources Utilization of Korean Immigrant Entrepreneurs in the Chicago Minority Area," in *Immigrants and Immigration Policy: Individual Skills, Family Ties, and Group Identities*, HO Duleep, PV Wunnava, eds. Greenwich, Conn.: JAI Press.

Kim, K. C., Hurh, W. M., & Fernandez, M. (1989). "Intra-group differences in business participation: three Asian immigrant groups." *The International migration review*, 23(1), 73–95.

Light, I. (1972) *Ethnic Enterprises in America: Business and Welfare Among Chinese, Japanese, and Blacks*. Berkeley, Calif: University of California Press.

Brent R. Moulton, "Random group effects and the precision of regression estimates," *Journal of Econometrics,* Volume 32, Issue 3, 1986, Pages 385–397, ISSN 0304-4076.

Portes A and Bach R. (1985) *Latin Journey: Cuban and Mexican Immigrants in the United States.* Berkeley, Calif.: University of California Press.

Waldinger R. (1986) *Through the Eye of the Needle: Immigrant Enterprise in New York's Garment Trades.* New York: New York University Press.

Waldinger R. (1989) "Structural Opportunity or Ethnic Advantage? Immigrant Business Development in New York." *International Migration Review* 23: 48–72.

PART IV

More Recent Cohorts

Part II documents that despite having schooling levels above those of U.S. natives, immigrant men from Asian developing countries in the post-1965 cohorts started their U.S. journeys with earnings substantially below those of U.S. natives. Yet, their earnings growth exceeded that of U.S. natives, leading to earnings convergence 10 to 15 years after immigration, as first predicted by Chiswick (1978, 1979, 1980). In Part IV, we explore whether these patterns persist for more recent cohorts.

References

Chiswick, B.R. (1978) "The Effect of Americanization on the Earnings of Foreign Born Men," *Journal of Political Economy*, October, pp. 897–922.

Chiswick, B.R. (1979) "The Economic Progress of Immigrants: Some Apparently Universal Patterns," in W. Fellner, ed., *Contemporary Economic Problems*. Washington, D.C.: American Enterprise Institute, pp. 359–99.

Chiswick, B.R. (1980) *An Analysis of the Economic Progress and Impact of Immigrants*, Department of Labor monograph, N.T.I.S. No. PB80-200454. Washington, DC: National Technical Information Service.

CHAPTER 17

Entry Earnings, Earnings Growth, and Human Capital Investment: The 1985–1990 and 1995–2000 Cohorts

Using microdata samples from the 1970 through 2000 censuses, Fig. 17.1 depicts the initial earnings of immigrant men (ages 25–64 in each census year) from the Asian developing countries as a percentage of the median earnings of U.S.-born men. The immigrant cohorts include men who came to the United States during the years 1965–1970 (for the 1970 census), 1975–1980 (for the 1980 census), 1985–1990 (for the 1990 census), and 1995–2000 (for the 2000 census). Annual earnings are measured for the year that precedes each census year.

As was true for the 1965–1970 and 1975–1980 cohorts, U.S.-born/foreign-born differences in schooling levels cannot explain the low entry earnings of the more recent cohorts relative to U.S. natives. In 1990, slightly more than 18% of U.S. natives, ages 25–64, had less than a high school degree versus 11% of Filipino immigrants, 19% of Chinese immigrants, 12% of Korean immigrants, and 12% of Indian immigrants. Fifty-one percent of U.S. natives had at least some college compared with 74% of Filipino immigrants, 70% of Chinese immigrants, 67% for Korean immigrants, and 78% for Indian immigrants (Table 17.1).[1] As was true for the earlier cohorts, large variations in English proficiency do not

---

[1] Based on the 1990 census (6% sample), the more detailed statistics for U.S.-born, 25–64 year old men, are: 4 years or less 1.19%; 5–8 years 4.37%; Some High School 12.58%; High School Graduate 30.83%; Some College 26.89%; Bachelors 15.28%; Graduate Degree 8.86%.

© The Author(s), under exclusive license to Springer Nature Switzerland AG 2020
H. Duleep et al., *Human Capital Investment*,
https://doi.org/10.1007/978-3-030-47083-8_17

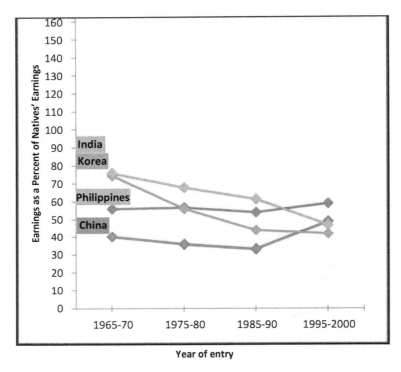

Fig. 17.1 Median earnings, the first five years, of immigrant men from the Asian developing countries as a percentage of U.S.-born men: The 1965–1970, 1975–1980, 1985–1990, and 1995–2000 cohorts. (Source: Authors' creation based on 1970, 1980, 1990, and 2000 Census PUMS data)

necessarily translate into earnings variations. The initial earnings of the 1995–2000 cohort of Chinese, Korean, and Indian immigrants are remarkably similar (Table 17.3) despite large differences in English proficiency (Table 17.2). Although the Asian developing-country immigrants had low initial earnings, Chap. 7 documented their high earnings growth; their earnings, ten years later, approached or exceeded the earnings of U.S. natives. Fig. 17.2 shows this to be an enduring pattern.[2]

Human capital investment occurs in many forms most of which are hard to measure; earnings growth embodies all aspects of human capital

---

[2] Also see Table 17.3.

Table 17.1 Entry education level of the 1995–2000 and 2005–2010 cohorts of 25–64-year-old male immigrants (percent)

	1985–1989 cohort, 1990 educational attainment						1995–2000 cohort, 2000 educational attainment					
	Less than high school	Some high school	High school degree	Some college	Bachelor's	Graduate degree	Less than high school	Some high school	High school degree	Some college	Bachelor's	Graduate degree
Philippines	5.9	4.7	15.6	28.2	37.0	8.7	5.7	2.5	21.3	20.1	43.5	6.9
China (all)	11.1	7.7	11.6	12.6	23.8	33.3	8.0	4.3	15.5	9.1	22.6	40.5
Mainland China							8.2	4.0	14.7	8.6	22.4	42.1
Hong Kong							5.7	7.6	25.6	16.4	25.6	19.1
Taiwan							1.5	0.6	11.6	14.9	31.7	39.8
Korea	5.7	6.2	21.5	19.1	23.0	24.5	1.9	1.1	19.6	14.5	35.2	27.6
India	4.3	8.0	9.4	12.1	26.3	39.9	2.1	2.9	7.9	4.5	42.4	40.2

Source: Authors' estimates. For the 1985–1990 cohort, estimates are based on a 6% microdata sample created by combining and reweighting the 1990 Census of Population Public Use 5% and 1% samples. For the 1995–2000 cohort, estimates are based on a 2000 Census of Population 6% microdata sample combining the 2000 Census of Population Public Use 5% and 1% samples. Appendix A provides information on sample sizes for entry cohorts

Notes: Foreign born are defined as persons born outside of the United States excluding those with U.S. parents. Groups are defined by country-of-origin information. Korea includes South, North, and unspecified Korea. The sample does not exclude students, the self-employed, and persons with zero earnings

Table 17.2 Initial levels of English proficiency for 25–64-year-old male immigrants who entered the United States in 1985–1990 and 1995–2000

	Percent who speak English poorly		Percent who speak English very well	
	1985–1990 entry cohort; 1990 data	1995–2000 entry cohort; 2000 data	1985–1990 entry cohort; 1990 data	1995–2000 entry cohort; 2000 data
Philippines	8.7	10.7	55.1	53.4
China (all)	38.2	34.8	23.4	28.3
Mainland China		35.4		27.3
Hong Kong		26.4		41.3
Taiwan		27.8		23.4
Korea	50.5	44.5	17.1	19.0
India	9.2	6.3	65.8	74.0

Source: Authors' estimates. For the 1985–1990 cohort, estimates are based on a 6% microdata sample created by combining and reweighting the 1990 Census of Population Public Use 5% and 1% samples. For the 1995–2000 cohort, estimates are based on a 2000 Census of Population 6% microdata sample combining the 2000 Census of Population Public Use 5% and 1% samples. Appendix A provides information on sample sizes for entry cohorts

Notes: Foreign born are defined as persons born outside of the United States excluding those with U.S. parents. Groups are defined by country-of-origin information. Korea includes South, North, and unspecified Korea. The sample does not exclude students, the self-employed, and persons with zero earnings. Speaks English Poorly includes those who speak English poorly or not at all; Speaks English Very Well includes those who speak English very well, or who only speak English

investment, measured and unmeasured. Nevertheless, we find from 2000 census data, that immigrant men in the Asian developing-country 1995–2000 entry cohorts are more likely to attend school than West European immigrants and U.S. natives (Fig. 17.3). A similar picture emerges with 2010 census data for the 2005–2010 cohort (Fig. 17.4).

China Disaggregated

We have throughout the book consolidated mainland China, Taiwan, and Hong Kong into one China. Disaggregating the 1995–2000 cohort reveals key differences among immigrants from these three places of origin.

China and Taiwan are similar in terms of the percentage of highly educated immigrants: for the cohort that entered the United States in 1995–2000, 42% of immigrants from mainland China and 40% of Taiwanese immigrants have a graduate degree, compared with 19% for

Fig. 17.2 Median earnings, the first five years and ten years later of immigrant men from the Asian developing countries as a percentage of U.S.-born men: the 1975–1980, 1985–1990, and 1995–2000 cohorts. (Source: Authors' creation based on 1980, 1990, 2000, and 2010 Census PUMS data)

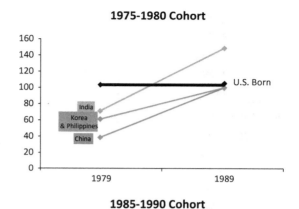

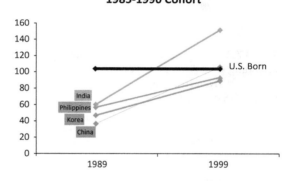

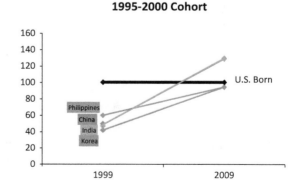

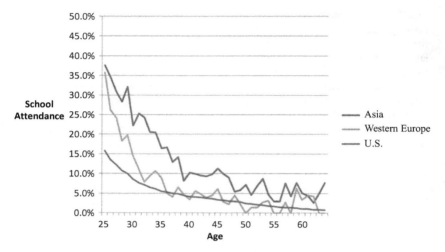

Fig. 17.3 Percentage of immigrant men from the Asian developing countries attending school by single years of age, compared with US natives and immigrant men from Western Europe, 2000. (Source: Authors' creation based on 2000 census PUMS data)

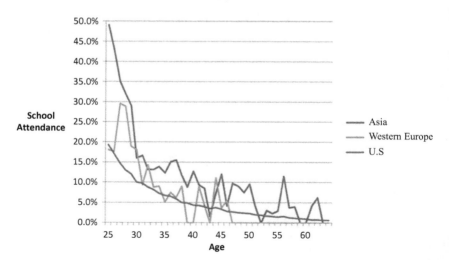

Fig. 17.4 Percentage of immigrant men from the Asian developing countries attending school by single years of age, compared with US natives and immigrant men from Western Europe, 2010. (Source: Authors' creation based on 2010 census PUMS data)

immigrants from Hong Kong (Table 17.1). At the opposite end of the educational spectrum, immigrants from mainland China and Hong Kong are similar with 12% and 13% lacking a high school degree versus 2% of Taiwanese immigrants.

Despite having lower levels of schooling than immigrants from mainland China and Taiwan, the entry earnings of immigrants from Hong Kong are relatively high perhaps reflecting their higher English proficiency.[3] Forty-one percent of recently arrived immigrants from Hong Kong report speaking English very well in 2000 versus 27% for immigrants from mainland China and 23% for immigrants from Taiwan (Table 17.2).

Resonating with a theme of an inverse relationship between adjusted entry earnings and earnings growth, relatively low earnings growth accompanies the high entry earnings of Hong Kong immigrants. Although they start at higher earnings than immigrants from mainland China, ten years later their earnings are almost the same (Table 17.3). In keeping with their relatively low earnings growth, immigrants from Hong Kong who have been in the United States five years or less are much less likely to attend school than immigrant men from mainland China and Taiwan.

ARE THE EARNINGS PATTERNS WE OBSERVE DUE TO ANTI-ASIAN DISCRIMINATION?

Looking at immigrant men entering within five years of the 1970–2000 censuses, we find that Asian immigrants from economically developing countries initially earn far less than U.S. natives, despite their higher levels of education, whereas West European and Canadian immigrants earn on a par or exceed the earnings of U.S. natives. We theorize that the differential earnings patterns stem from differences in skill transferability. Persons migrating from developing countries may have greater problems transferring skills due to greater differences in how labor is used to produce goods and services. They may also be more willing than those in more developed economies to accept skill transferability problems and hope to gain from migration even if it means switching careers. Persons in economically developing countries will be less likely to do this. But, could not the patterns we observe be due to discrimination?

---

[3] An interesting further analysis would be to see if the Hong Kong earnings profile of relatively high initial earnings persists when the sample is divided by level of English proficiency.

Table 17.3 Entry earnings and earnings after ten years for immigrant men (Ages 25–54) who entered the US in 1975–1980, 1985–1990, and in 1995–2000 as a percentage of U.S.-born earnings

	1985–1990 entry cohort			1995–2000 entry cohort		
	At entry: 1989 earnings	10 years later: 1999 earnings	10-year growth rate	At entry: 1999 earnings	10 years later: 2009 earnings	10-year growth rate
	Ages 25–54	Ages 35–64		Ages 25–54	Ages 35–64	
Philippines	54%	89%	55.5	59%	93%	39.4
China (all)	35%	102%	175.5	49%	129%	133.3
Mainland China				47%	129%	143.9
Hong Kong				74%	132%	57.8
Taiwan				36%	114%	181.5
Korea	45%	86%	80.3	42%	94%	98.9
India	58%	146%	138.0	47%	129%	143.9

Source: Authors' estimates. For the 1985–1990 cohort, entry earnings estimates are based on a 6% microdata sample combining the 1990 Census of Population Public use 5% and 1% samples; earnings 10 years later are based on a 6% sample combining the 2000 Census of Population 5% and 1% samples. For the 1995–2000 cohort, entry earnings estimates are based on a 6% microdata sample combining the 2000 Census of Population 5% and 1% samples; earnings ten years later, for the 1995–2000 cohort, are based on a 3% microdata sample combining the American Community Service 2008–2010 1% files. The estimates of immigrant earnings ten years later are likely underestimates because earnings are being measured eight to ten years later, not ten years later. We encourage replication particularly since as of this writing, there is a 10% 2010 file. Appendix A provides information on sample sizes for entry cohorts at entry and ten years later

Notes: Foreign born are defined as persons born outside of the United States excluding those with U.S. parents. Groups are defined by country-of-origin information. Korea includes South, North, and unspecified Korea. The sample does not exclude students, the self-employed, and persons with zero earnings

As described in Chap. 2, Asian immigrants in America experienced grave discrimination in the nineteenth and early twentieth centuries. Comparing the 1960 earnings of U.S.-born Asians and whites adjusting for earnings-related characteristics such as years of schooling and work experience, Duleep and Sanders (2012) found that U.S.-born Japanese men earned 23% less than non-Hispanic whites with average Japanese characteristics, U.S.-born Chinese men earned 13% less than comparable whites, and U.S.-born Filipino men earned 39% less than whites with average Filipino characteristics. Furthermore, the 1960 earnings gap persists taking into account every conceivable possible explanation for it, other than labor market discrimination.

By 1980, there was near closure of the unexplained Asian/white earnings gap for the U.S. born. This does not mean that U.S.-born Asians stopped facing labor market discrimination. Indeed, Duleep and Sanders (1992) found that within industries, U.S.-born men in all Asian groups were less likely to attain management positions than U.S.-born non-Hispanic white men. Minorities respond to discrimination by choosing paths that are less impacted by discrimination, and this affects statistical measurements of discrimination (Duleep and Zalokar, 1991). Given this, how can we argue that skill transferability linked to immigration from economically developing countries underlies the earnings patterns that we have described, instead of a story of labor market discrimination?

We first note the constancy over time in our findings. For all of the cohorts we have studied, from the 1965–1970 entry cohort to the 1995–2000 entry cohort, we find that Asian immigrant men from economically developing countries have low initial earnings followed by high earnings growth. We find the same pattern even though statistical measures of anti-Asian discrimination decrease over time. Moreover, men from Eastern Europe follow the same pattern as do Asian immigrant men from economically developing countries. Finally, we note that men from Japan have earnings trajectories resembling immigrant men from Western Europe and Canada: they have high initial earnings and low earnings growth. Whatever the effect of reduced discrimination over these decades, the patterns we observe seem better explained by immigrant skill transferability issues rather than ethnicity.

REFERENCES

Duleep, Harriet and Nadja Zalokar, "The Measurement of Labor Market Discrimination When Minorities Respond to Discrimination," in Richard Cornwall and Phanindra Wunnava, eds., *New Approaches to Economic and Social Analyses of Discrimination*, NY: Praeger Press, 1991

Duleep, Harriet and Seth Sanders, 1992, "An Exploratory Analysis of Discrimination at the Top: American-Born Asian and White Men," *Industrial Relations*, Fall 1992

Duleep, Harriet and Seth Sanders, 2012, "The Economic Status of Asian Americans Before and After the Civil Rights Act," IZA Discussion Paper No. 6639

PART V

The Impact of Refugee Status

Asian immigration to the U.S. changed abruptly in 1978. From 1970 to 1977, the Philippines, China, Korea, India, and Japan led U.S. Asian immigration. As the 1970s came to a close, Cambodia, Laos, and, above all, Vietnam emerged as key contributors. Indeed, from 1978 to 1980, Vietnam was *the* leading source of Asian immigration. For the countries that dominated post-1965 Asian immigration, and were the focus of Parts II through IV, a quarter to a third of their 1970–1980 immigration occurred after 1977 compared with 86% of Vietnamese, 94% of Cambodian, and 97% of Laotian immigration.

If nineteenth-century and early twentieth-century Asian migration was America's first wave of Asian immigration, and post-1965 immigration its second wave, then the refugee influx that began with the end of the Vietnam War is the third wave, distinguished from its predecessors by source region and type of admission. The Asian groups we studied in Parts II through IV included few refugees: 4% of Chinese immigrants, less than 0.2% of Filipino immigrants, 0.03% of Korean immigrants, and less than 0.02% of Indian immigrants in 1980 were refugees. In contrast, of those given permanent visas in 1980, 90% of Vietnamese, 95% of Cambodian, and 99% of Laotian immigrants were refugees.

The experiences of these groups shed light on a frequently voiced concern in today's debates about immigration—how do poorly educated immigrants fare in today's economies? Though the first entrants from Vietnam were well educated, their successors were not. Cambodians and Laotians had even less formal schooling.

Prior chapters showed that the earnings patterns of immigrants in the Asian economically developing groups resembled each other but differed dramatically from the earnings trajectories of West European, Canadian, and Japanese immigrants. Part V compares the assimilation patterns of Indochinese entrants with the Asian developing-country groups of Parts II through IV. To facilitate this comparison, we aggregate the previously studied immigrants from the Philippines, China, Korea, and India into one comparison group.

Chapter 18 highlights factors that may affect the economic assimilation of refugees. Using 1980–2000 census data, Chap. 19 follows the earnings and human-capital investment of Southeast Asian refugee men who came to the United States between 1975 and 1980. Chapter 20 explores how South East Asian refugee women aid family economic assimilation. Part V concludes with a focus on the 1985–1990 cohort of Indochinese refugees.[1]

References

Wright, Mary Bowen, 1980. "Indochinese," in Stephan Thernstrom, ed. *Harvard Encyclopedia of American Ethnic Groups*. Harvard University Press, Cambridge, MA and London, England, pp. 508–513.

Immigration and Naturalization Service. 1980. *Statistical Yearbook of the Immigration and Naturalization Service*. Washington, D.C.: U.S. Government Printing Office.

---

[1] Refer to Bowen (1980).

CHAPTER 18

Factors Associated with Refugee Status

The reasons immigrants come to the United States and the processes through which they enter can affect their characteristics and their earnings and labor market behavior. South East Asian refugees differ in several ways from the immigrants who hail from the economically developing Asian countries studied in Parts II through IV.[1]

COMMUNITY TIES AND EXTENDED FAMILY

The vast majority of Indochinese were admitted to the United States as refugees and not via the skill or kinship-based entry programs discussed in Chap. 1. As refugees, there were no guarantees that they would have either a U.S. employer desiring their skills or the advantage of a U.S.-settled family.[2] The earliest cohorts also lacked existing co-ethnic enclaves to provide information about their adopted country, jobs, or other support. Moreover, the U.S. government dispersed the first Indochinese refugees across the nation.[3] These efforts were ultimately unsuccessful, as each scattered group migrated to identifiable areas of ethnic concentration.

---

[1] Please refer to Reimers (2005, pp. 276–288).

[2] Many of the later Indochinese entry cohorts did, of course, have family ties in the U.S. that facilitated, but was not legally required for admission to the U.S. from a refugee camp.

[3] The dispersal policy established by the U.S. for the resettlement of Indochinese refugees was motivated by two concerns: the impact on local communities and the geographic distribution of volunteer refugee assistance organizations. Reimers (2005, p. 281) writes:

© The Author(s), under exclusive license to Springer Nature Switzerland AG 2020
H. Duleep et al., *Human Capital Investment*,
https://doi.org/10.1007/978-3-030-47083-8_18

Table 18.1 US geographic distribution of immigrant families who have been in the US five years or less in 1980 and in 1990

US Region	Western Europe, 1975–80 cohort, 1980	Thailand 1975–80 cohort, 1980	Thailand 1985–90 cohort, 1990	Vietnam 1975–80 cohort, 1980	Vietnam 1985–90 cohort, 1990	Cambodia 1975–80 cohort, 1980	Cambodia 1985–90 cohort, 1990	Laos 1975–80 cohort, 1980	Laos 1985–90 cohort, 1990
California	0.14	0.38	0.38	0.37	0.55	0.39	0.54	0.21	0.45
Hawaii	0.00	0.00	0.00	0.01	0.01	0.00	0.00	0.03	0.01
Other Western State	0.03	0.16	0.20	0.09	0.07	0.25	0.13	0.16	0.06
North Central	0.10	0.15	0.08	0.12	0.05	0.12	0.05	0.30	0.25
South	0.17	0.21	0.22	0.33	0.20	0.16	0.12	0.20	0.13
East	0.56	0.10	0.12	0.08	0.12	0.08	0.16	0.10	0.10

Source: Authors' estimates based on 1980 and 1990 Census 5% Public Use Samples

Table 18.1 compares the initial geographic locations of the 1975–1980 and 1985–1990 cohorts of Indochinese refugee families with that of Thai families, who were not predominantly refugee. Although there was little inter-cohort change in the geographic distribution of entering Thai families, the distribution of entering Indochinese changed markedly. For all three refugee groups, the 1985–1990 cohort was much more likely to settle in California upon first arrival than the 1975–1980 cohort, the cohort most subject to the U.S. refugee dispersion policy. Table 18.2 follows the geographic location of the 1975–1980 cohort of Indochinese refugees during their first 10–15 years in the United States. With time in the United States, these refugees migrated from North Central, Southern, and Eastern United States, where the resettlement policy first encouraged them to live, to California. Their ultimate settlement pattern resembles the initial settlement pattern of the 1985–1990 cohort.

"American officials feared that the development of ethnic enclaves with distinct cultures would lead to cultural conflict and have an adverse economic impact on local host communities. The government wanted to 'avoid another Miami,' as one authority put it. Officials therefore pursued a policy of dispersal of the refugees throughout the United States." Over time, the impact of the dispersal policy decreased, as later entering cohorts were more likely to have family already in the U.S., or were otherwise able to find sponsors within their own immigrant community. Refer to Hein (1995, pp. 50–54) for a discussion of resettlement policies.

Table 18.2 US geographic distribution in 1980 and 1990 of Indochinese families who immigrated between 1975 and 1980

	Vietnam		Cambodia		Laos	
	1980	1990	1980	1990	1980	1990
California	0.37	0.47	0.39	0.47	0.21	0.43
Hawaii	0.01	0.01	0.00	0.00	0.03	0.00
Other Western State	0.09	0.08	0.25	0.09	0.16	0.12
North Central	0.12	0.08	0.12	0.10	0.30	0.19
South	0.33	0.31	0.16	0.22	0.20	0.20
East	0.08	0.05	0.08	0.12	0.10	0.06

Source: Authors' estimates based on 1980 and 1990 Census 5% Public Use Samples

The federally encouraged dispersion of Indochinese refugees may have inadvertently slowed their economic assimilation by delaying the formation of immigrant networks. Chapter 16 suggests that extended family and community networks promote human capital investment among immigrants who have low initial levels of U.S.-specific skills[4]: sibling admissions positively correlate with both earnings growth and self-employment. The more limited initial presence of extended family and slower growth of enclaves may have dampened the human capital investment and earnings growth of Indochinese refugees compared with non-refugee immigrants with similar entry-level skills. On the other hand, federal refugee assistance programs and an immediate eligibility for food stamps, Medicare and other support programs—assistance that other immigrants must wait to become eligible for—may have helped investment in human capital by facilitating further schooling.

---

[4] For case study evidence see, for instance, Bailey (1987), Bonacich and Modell (1980), Fawcett, Carino, Park, and Gardner (1990), Gallo and Bailey (1996), Jiobu (1996), Light (1972), Waldinger (1986, 1989), Khandelwal (1996), and Kim and Hurh (1996). Gallo and Bailey (1996, p. 208) comment: "Comparisons between refugees and other immigrants also suggest the importance of networks. Refugees tend to arrive without connections and research suggests that their earnings are lower and grow more slowly than non-refugee immigrants with similar characteristics. Indeed, the government has tried to provide formal job finding and training assistance in lieu of the informal sources of information embodied in networks. But a wide variety of research suggests that in the U.S., formal labor market intermediaries are much less effective than informal networks."

Selection on Observed and Unobserved Characteristics

Refugees are subject to particular forms of self-selection that may affect observed and unobserved human capital characteristics. Early refugees from Cuba's communist government, for example, were part of Cuba's middle and upper classes—the groups most persecuted by the government—and thus were well educated.[5]

In Vietnam, personal or family association with the former South Vietnamese government or its military crossed class lines. In Cambodia, the campaign of mass-collectivization aimed at city dwellers and the educated expanded across class lines. In Laos, an ethnic dimension colored the war. Most of the poorly educated and largely rural Hmong backed the Royal Lao government, taking up arms with covert U.S. military support against the communist Pathet Lao and North Vietnamese forces based in Laos. The educational mix of South East Asian refugees resulting from these diverse forces is displayed in Table 18.3.

Compared with working-age Asian immigrant men from developing countries, a high percentage of entering Indochinese refugees possessed only a grade school education; a much lower percentage possessed postsecondary degrees. Important intergroup variations are evident. Cambodians and Laotians have lower levels of schooling than the Vietnamese. Dividing by ethnicity increases the variance (Table 18.3). Over two-thirds of the ethnic Vietnamese report at least a high school degree. For the Hmong, nearly half of adult working-age men report four years or less of schooling.

The different forms of self-selection for refugees than other immigrants may affect less tangible characteristics, such as motivation and unmeasured ability. Common to many models of domestic and international migration (Sjaastad 1962; Polachek and Horvath 1977; Chiswick 1978) is the notion that migration is an investment that involves both direct and opportunity costs: Individuals or families will migrate only when the returns to migration exceed these costs.

> ... if greater labor market ability and motivation raise earnings relatively more than they raise the cost of migration, the rate of return from migration

---

[5] See Borjas (1990, pp. 128–129) for a discussion of skill level of various waves of Cuban immigration to the U.S.

Table 18.3 Distribution of 25–64-year-old male Indochinese by education level on 1990 census (percent)

	4 years or less	5–8 years	Some high school	High school	Some college	Bachelors	Graduate degree
Place of birth in Indochina:							
Vietnam	10.6	8.3	18.4	17.8	28.2	11.2	5.5
Cambodia	28.1	10.8	15.0	15.0	22.2	6.3	2.6
Laos	30.4	11.0	13.4	18.0	20.2	4.8	2.3
Selected ethnic groups in Indochina:							
Vietnamese	8.9	7.0	17.4	18.2	29.5	12.7	6.4
Cambodian	29.1	10.8	13.9	14.9	22.3	6.5	2.6
Chinese	16.9	13.5	22.2	17.1	22.0	5.5	2.9
Laotian	26.3	11.9	14.6	20.3	20.7	4.7	2.6
Hmong	45.1	10.6	9.6	9.8	19.7	3.9	1.3
Comparison groups							
China, India, Korea, Philippines combined	3.4	3.8	6.1	12.8	17.7	26.6	29.6
Western Europe	9.6	14.5	11.8	17.8	16.9	11.1	18.3

Source: Authors' estimates based on a 6% microdata sample created by combining and reweighting the 1990 Census of Population Public Use 5% and 1% samples. The ethnicities shown here are based upon answers to the first ancestry code on the census

is greater for the more able and motivated, and they will have a higher propensity to migrate.[6] (Chiswick 1978, pp. 900–901)

Yet, refugees experience much less personal choice in the decision to migrate.

Although economic migrants tend to be favorably self-selected on the basis of high innate ability and economic motivation, this self-selection is less intense for refugees with similar demographic characteristics, whose migration is primarily influenced by the political and social environment. (Chiswick 1979)

---

[6] Other relevant studies on the relationship between the propensity to migrate and earnings ability include O'Neill (1970), Schultz (1975), Schwartz (1976), and Yezer and Thurston (1976).

This line of reasoning suggests that, among refugees, the less the personal choice affects the decision to migrate, the less the positive selection on variables that engender economic success.

Indochinese refugees left their countries under varied circumstances. The first were the Vietnamese who were evacuated by the United States or left Vietnam with more conventional transportation shortly before or after the fall of Saigon in 1975. Later, Vietnamese refugees often took great personal risks with reports of as many as half dying at sea as they escaped in small, unsafe boats.[7]

Little choice also colored the migration of Cambodian and Laotian refugees—crossing into Thailand, fleeing mass killings (as in the case of the Hmong and the early Cambodian refugees), or ordered across the border at gun point, as with the Cambodian captives of the Khmer Rouge forces that fled the 1979 Vietnamese invasion of Cambodia. Although one cannot link type of migration with amorphous characteristics such as ability and motivation, disability rates (including their duration over time) may gauge the physical and emotional trauma associated with various migrations.

For working-age immigrant men in the 1975–1980 cohort, about 5% of the Indochinese reported a work-inhibiting disability in 1980 versus 3.8% of immigrants from the big four Asian developing countries and 2.1% of West European immigrants. Among the Indochinese, 3.9% of the ethnic Vietnamese versus 11.3% of the Hmong reported a disability (Table 18.4).

Disability rates tend to increase with age. As this occurs, the difference between the disability rates of the 1975–1980 cohort of Indochinese and that of other immigrants falls. At ages 35–64 in 1990, the Indochinese disability rate for the 1975–1980 cohort is 8.7% versus 6.3% for West European immigrant men. The Vietnamese disability rate is about the same as the West European rate. The Hmong are the exception. After 10 to 15 years in the United States, their disability rate nearly tripled to more than 30%.

---

[7] Beyond the migration decision, Caplan, Choy, and Whitmore (1991, p. 44) raise the issue of selective survival. Citing reports that as many as half of the boat people died at sea, they speculate that "it is possible that pre-selection in conjunction with survival factors may have resulted in bias towards resettlement success for those who made it to the U.S."

Table 18.4 Disability rates in 1980 of the 1975–1980 entry cohort: immigrant men, 25–54 years old (percent)

Place of birth in Indochina:	
Vietnam, Cambodia and Laos	5.1
Vietnam	4.5
Cambodia	5.6
Laos	7.8
Ethnicity (born in Indochina):	
Vietnamese	3.9
Cambodian	5.7
Chinese	6.5
Laotian	7.4
Hmong	11.3
Non-South East Asian comparison groups	
China, India, Korea, Philippines	3.8
Western Europe	2.1

Source: Authors' estimates based on the 1980 Census of Population, 5% "A" Public Use Sample. The ethnicities shown here are based upon answers to the first ancestry code on the census

Skill Transferability

Source-country to host-country skill transferability may also differ between refugees and other immigrants.

> Since the earning power of one's skills plays a primary role in economic migration and a secondary role in refugee migration, a cohort of the latter is likely to include a larger proportion of workers with skills that have little international transferability. Refugee migration generally arises from a sudden or unexpected change in political conditions,... refugees are less likely than economic migrants to have acquired readily transferable skills and are more likely to have made investments specific to their country of origin. (Chiswick 1979, pp. 365–366)

To the extent that skill transferability is lower for Southeast Asian refugees, as compared to immigrants from the Philippines, China, Korea, and India, we would expect holding level of source-country human capital constant, greater investment in U.S. human capital, hence higher earnings growth.

Permanence

A key component of the IHCI model is permanence. Refugees who cling to the prospect of returning home once a war or other turmoil has ended may invest less in U.S.-specific human capital than those for whom returning home is a lost hope.

Chapter 10 approximated relative emigration rates by measuring changes in group sizes on the 1980 and 1990 census samples for the 1975–1980 entry cohort.[8] Over the ten years between censuses, the weighted count of Indochinese men from the 1975 to 1980 cohort fell by only 9%, with a fall of 8% for men who were 25–39 years old in 1980 and a fall of 11.5% for men who were 40–54 years old. These rates resemble the drop in cohort size for Asian immigrants from developing countries (Chap. 10).[9] Indochinese refugees, like the Asian developing country immigrants, are highly permanent.

We theorized and provided supporting empirical evidence that the decisions to migrate, to invest, and to stay in the host country are jointly determined. Here, an external situation—the war in Indochina—determined migration and permanence.

Concluding Remarks

All elements in the Immigrant Human Capital Investment model that would increase human capital investment apply to Southeast Asian refugees. Most would have had difficulty transferring their occupational skills to the United States. The incentives for human capital investment are all the greater given their U.S. permanence. Working in the opposite direction are the potentially negative effects of forced migration, poor health,

---

[8] Emigration rates may be an unsatisfactory measure of the initial intent to stay or leave since changes over time may alter initial intentions. For instance, we hypothesized earlier that an improving Korean economy may have lured back Korea immigrants who had initially planned on staying. Changes may also occur in the extent to which skills learned in the U.S. are applicable to the home country. Thus, the rise of the computer industry in India may have encouraged a reverse migration.

[9] There are several limitations of this method of measuring emigration rates. In addition to emigration, changes in the count could arise from a number of other factors: group-specific differences in response rates to the 1980 and 1990 censuses, confusion about the place-of-birth question on the census with a country-of-citizenship question, sampling error from the 5% and 6% samples used, the inclusion of those who entered between January 1, 1980, and April 15, 1980, in the 1980 count for this cohort, and mortality.

emotional trauma, and the initial absence of immigrant enclaves and family ties. Against this backdrop, the next two chapters examine the earnings and labor force behavior of the first entrants of Indochinese refugee men and women who came to the United States between 1975 and 1980.

REFERENCES

Bailey, Thomas R. (1987) *Immigrant and Native Workers: Contrasts and Competition*, Conservation of Human Resources, Boulder and London: Westview Press.

Bonacich E, Modell J. 1980. *The Economic Basis of Ethnic Solidarity: Small Business in the Japanese-American Community*. Berkeley, Calif: University of California Press.

Borjas, George J. *Friends or Strangers: the Impact of Immigrants on the U.S. Economy*. New York: Basic Books, 1990.

Caplan, N., Choy, M. H., & Whitmore, J. K. (1991). *Children of the boat people: A study of educational success*. Ann Arbor: University of Michigan Press.

Chiswick, B.R. (1978) "The Effect of Americanization on the Earnings of Foreign-Born Men," *Journal of Political Economy*, October, pp. 897–922.

Chiswick, B.R. (1979) "The Economic Progress of Immigrants: Some Apparently Universal Patterns," in W. Fellner, ed., *Contemporary Economic Problems*. Washington, D.C.: American Enterprise Institute, pp. 359–99.

Fawcett JT, Carino BV, Park IH, Gardner RW. 1990. "Selectivity and Diversity: The Effects of U.S. Immigration Policy on Immigrant Characteristics." Paper presented at the annual meeting of the Population Association of America.

Gallo C, Bailey TR. (1996) "Social Networks and Skills-Based Immigration Policy," in *Immigrants and Immigration Policy: Individual Skills, Family Ties, and Group Identities*, HO Duleep, PV Wunnava, eds. Greenwich, Conn.: JAI Press.

Hein J. 1995. *From Vietnam, Laos, and Cambodia: A Refugee Experience in the United States*. New York: Twayne Publishers.

Jiobu RM. 1996. "Explaining the Ethnic Effect," in *Immigrants and Immigration Policy: Individual Skills, Family Ties, and Group Identities*, HO Duleep, PV Wunnava, eds. Greenwich, Conn.: JAI Press.

Khandelwal M. 1996. "Indian Networks in the United States: Class and Transnational Identities," in *Immigrants and Immigration Policy: Individual Skills, Family Ties, and Group Identities*, HO Duleep, PV Wunnava, eds. Greenwich, Conn.: JAI Press.

Kim KC and Hurh WM. 1996. "Ethnic Resources Utilization of Korean Immigrant Entrepreneurs in the Chicago Minority Area," in *Immigrants and Immigration Policy: Individual Skills, Family Ties, and Group Identities*, HO Duleep, PV Wunnava, eds. Greenwich, Conn.: JAI Press.

Light, I. (1972) *Ethnic Enterprises in America: Business and Welfare Among Chinese, Japanese, and Blacks.* Berkeley, Calif: University of California Press.

O'Neill, June A., *The Effect of Income and Education on Inter-Regional Migration*, Ph.D. dissertation, Department of Economics, Columbia University, 1970.

Polachek, Solomon W., and Francis W. Horvath, "A Life Cycle Approach to Migration: Analysis of the Perspicacious Peregrinator," in R. G. Ehrenberg (ed.), *Research in Labor Economics*, Vol. 1 (Greenwich, Ct: JAI Press, Inc., 1977), 103–150.

Reimers David M. 2005. *Other Immigrants: The Global Origins of the American People*, NY: NYU Press.

Schultz, Theodore W., "The Value of the Ability to Deal with Disequilibria," *Journal of Economic Literature*, September, 1975, PP. 827–46.

Schwartz, Abba, "Migration, Age and Education," *Journal of Political Economy*, August 1976, pp. 701–719.

Sjaastad, L. (1962). "The costs and returns of human migration," *Journal of Political Economy* 70: 80–93.

Waldinger R. (1986) *Through the Eye of the Needle: Immigrant Enterprise in New York's Garment Trades.* New York: New York University Press.

Waldinger R. (1989) "Structural Opportunity or Ethnic Advantage? Immigrant Business Development in New York." *International Migration Review* 23: 48–72.

Yezer AMJ, Thurston L. 1976. "Migration Patterns and Income Change: Implications for the Human Capital Approach to Migration." *Southern Economic Journal* 42 (4): 693–702.

CHAPTER 19

The Earnings and Human Capital Investment of Southeast Asian Refugee Men: The 1975–1980 Cohort

Indochinese refugee men who entered the United States between 1975 and 1980 began their New World lives at much lower earnings than the developing-country Asian immigrants studied in Parts II–IV. Their 1979 median earnings are 38% of U.S.-born natives' earnings versus 53% for the developing-country Asian immigrants (Table 19.1). Despite lower education levels, their earnings increase 97% from 1979 to 1989, reaching 77% of the median earnings for U.S. natives of the same age. The largest of the Indochinese ethnic groups, the Vietnamese, reached 92% of the median earnings of U.S.-born men. The median earnings of the developing country Asian immigrants (immigrants from the Philippines, China, Korea, and India) increased 96%, reaching 108% of U.S.-born earnings.

We have often used the median to measure the earnings growth of an average immigrant. The median has many advantages. It is less sensitive than the mean to small sample sizes, unaffected by the truncation of earnings in the census, and it circumvents "the need" to exclude the self-employed. However, the large number of zero earners in 1979 among the Indochinese yields fantastic growth rates for some groups. With median earnings of only $55 in 1979, refugee men from Laos show a ten-year growth of 21,918%! Moreover, with median earnings of zero in each period, the median suggests zero improvement for the Hmong.

Nevertheless, the median estimates are insightful. For most Indochinese groups, large changes in median earnings reflect large changes in the

Table 19.1 Entry earnings and Earnings after ten years, measured at the median, for men (ages 25–54) who entered the U.S. in 1975–1980 1989 dollars (percentage of U.S. native born earnings in parentheses)

Place of birth in Indochina	1979 earnings ages 25–54	1989 earnings ages 35–64	10-year growth rate
Vietnam, Cambodia and Laos	10,247 (38%)	20,143 (77%)	96.6
Vietnam	13,250 (49%)	23,106 (89%)	74.4
Cambodia	3208 (12%)	16,739 (64%)	421.8
Laos	55 (0.2%)	12,110 (47%)	21,918.2
Ethnicity (born in Indochina)			
Vietnamese	15,093 (56%)	23,857 (92%)	58.1
Cambodian	3711 (14%)	16,167 (62%)	335.7
Chinese	9223 (34%)	22,256 (86%)	141.3
Laotian	512 (2%)	13,972 (54%)	2628.9
Hmong	0 (0%)	0 (0%)	–
China, India, Korea, Philippines	14,342 (53%)	28,167 (108%)	96.4

Source: Authors' estimates based on a 6% 1980 file created by combining and reweighting the 5% and 1% files of the 1980 Census of Population and a 6% microdata 1990 sample created by combining and reweighting the 1990 Census of Population Public Use 5% and 1% samples. Appendix A provides information on sample sizes for entry cohorts at entry and ten years later

Notes: Native born are defined as persons born in the U.S.; foreign born are defined as persons born outside of the United States excluding those with U.S. parents. Groups are defined by country-of-origin information. The ethnicities shown here are based upon answers to the first ancestry code on the census. The sample does not exclude students, the self-employed, and persons with zero earnings

percentage of the group employed. To supplement our analysis of earnings at the median, we use changes in mean earnings. The ten-year growth rate for all Indochinese refugee men, measured at the mean, remains a relatively high 93% (Table 19.2).

The Immigrant Human Capital Investment (IHCI) Model does not predict that immigrant groups with low initial earnings will necessarily have greater earnings growth than groups with high initial earnings. It predicts that this will occur when the groups begin their U.S. journeys with similar levels of source-country human capital. Among the Indochinese, the non-Vietnamese ethnic groups—who have lower initial earnings than the Vietnamese—have relatively high earnings growth. From 1979 to 1989, the earnings of Cambodian, Chinese, and Laotian men grew 125, 120, and 105%, respectively, versus 81% for Vietnamese men. The earnings growth of these groups approaches or exceeds that of

Table 19.2 Entry earnings and earnings after ten years, measured at the mean, for men (ages 25–54) who entered the US in 1975–1980 1989 dollars (percentage of US native born earnings in parentheses)

Place of birth in French Indochina	1979 earnings ages 25–54	1989 earnings ages 35–64	10-year growth rate
Vietnam, Cambodia and Laos	12,305 (42%)	23,711 (76%)	92.7
Vietnam	13,790 (47%)	26,397 (85%)	91.4
Cambodia	7783 (27%)	19,127 (61%)	145.8
Laos	6756 (23%)	12,412 (40%)	83.7
Ethnicity			
Vietnamese	14,827 (51%)	26,781 (86%)	80.6
Cambodian	8508 (29%)	19,137 (61%)	124.9
Chinese	13,194 (45%)	29,098 (93%)	120.5
Laotian	6778 (23%)	13,925 (45%)	105.4
Hmong	5843 (20%)	7637 (24%)	30.7
China, India, Korea, Philippines	17,508 (60%)	36,436 (117%)	108.1
U.S.-born	29,308	31,140	6.3

Source: Authors' estimates based on a 6% 1980 file created by combining and reweighting the 5% and 1% files of the 1980 Census of Population and a 6% microdata 1990 sample created by combining and reweighting the 1990 Census of Population Public Use 5% and 1% samples. Appendix A provides information on sample sizes for entry cohorts at entry and ten years later

Notes: Native born are defined as persons born in the U.S.; foreign born are defined as persons born outside of the United States excluding those with U.S. parents. Groups are defined by country-of-origin information. The ethnicities shown here are based upon answers to the first ancestry code on the census. The sample does not exclude students, the self-employed, and persons with zero earnings

the developing-country Asian immigrants, whose mean earnings grew 108%. The higher earnings growth of the Cambodians and Laotians is notable given their much lower education levels than the Vietnamese (Table 21.1): nearly 60% of Cambodian and 62% of Laotian men have less than twelve years of schooling versus 23% of the Vietnamese.

In contrast to the inverse relationship between entry earnings and earnings growth that has often appeared in this book (even without controlling for intergroup differences in education), the Hmong experience low initial earnings and low earnings growth. Measured at the mean, earnings did grow for the Hmong, but at a very slow rate: their earnings as a percentage of U.S. natives increased 4 percentage points in ten years, from 20% to 24% of U.S. natives' earnings (Table 19.2).

Table 19.3 Comparison of entry earnings and ten-year real earnings growth rates by age and education for 1975–1980 Indochinese and comparison immigrant groups, 1989 dollars

Age and schooling levels	Vietnam, Cambodia and Laos			China, India, Korea, Philippines		
	Entry earnings 1979	Earnings after 10 years	Earnings growth	Entry earnings 1979	Earnings after 10 years	Earnings growth
25–39 years old; 1–12 years of school	5076	17,300	241	12,678	19,545	54
25–39 years old; more than 12 years of school	16,902	28,370	68	15,212	35,000	130
40–54 years old; 1–12 years of school	2373	10,320	335	11,834	14,100	19
40–54 years old; more than 12 years of school	18,591	25,693	38	16,902	24,000	42

Source: Authors' estimates based on a 6% 1980 file created by combining and reweighting the 5% and 1% files of the 1980 Census of Population and a 6% microdata 1990 sample created by combining and reweighting the 1990 Census of Population Public Use 5% and 1% samples. Appendix A provides information on sample sizes for entry cohorts at entry and ten years later

Notes: Native born are defined as persons born in the U.S.; foreign born are defined as persons born outside of the United States excluding those with U.S. parents. Groups are defined by country-of-origin information. The ethnicities shown here are based upon answers to the first ancestry code on the census. The sample does not exclude students, the self-employed, and persons with zero earnings

To partially control for differences in human capital levels, we measure ten-year changes in median earnings within four age/education groups (Table 19.3). Among the more educated, who were predominantly Vietnamese, the Indochinese begin their U.S. lives at *higher* earnings than the developing-country Asian immigrants, and they experience lower earnings growth. Among the less educated, the Indochinese begin their U.S. lives at lower earnings than the developing-country Asian immigrants, and they experience higher earnings growth.

Finally, we measure earnings 20 years later for the 1975–1980 cohort (Table 19.4 and Fig. 19.1). As one would expect, progress toward

Table 19.4 Median earnings of the 1975–1980 cohort during the first 10 years and 20 years later (earnings as a percent of median earnings of U.S. native-born men are in parentheses)

	Earnings measured at the median					
Place of birth in Indochina	1979 Earnings, Ages 25–44	1989 Earnings, Ages 35–54	1999 Earnings, Ages 45–64	20-Year Growth Rate	Change in percentage points 1979 to 1989	1989 to 1999
Indochina	10,247 (40%)	21,667 (77%)	19,220 (81%)	87.6	37	4
Vietnam	13,660 (53%)	25,000 (89%)	22,140 (93%)	62.1	36	4
Cambodia	3421 (13%)	17,897 (64%)	16,864 (71%)	393.0	51	7
Laos	0 (0%)	12,516 (45%)	14,880 (63%)	–	45	18
Ethnicity (born in Indochina)						
Vietnamese	15,366 (60%)	25,553 (91%)	22,940 (96%)	49.3	31	5
Cambodian	3711 (15%)	18,260 (65%)	16,095 (68%)	333.7	50	3
Chinese	9223 (36%)	25,167 (90%)	22,134 (93%)	140.0	54	3
Laotian	691 (3%)	15,133 (54%)	14,942 (63%)	2062.4	51	9
Hmong	– (0%)	– (0%)	4526 (19%)	–	0	19
China, India, Korea, Philippines	14,854 (58%)	30,000 (107%)	27,187 (114%)	83.0	49	7

Source: Authors' estimates based on a 6% 1980 file created by combining and reweighting the 5% and 1% files of the 1980 Census of Population and a 6% microdata 1990 sample created by combining and reweighting the 1990 Census of Population Public Use 5% and 1% samples. Appendix A provides information on sample sizes for entry cohorts at entry and ten years later

Notes: Native born are defined as persons born in the U.S.; foreign born are defined as persons born outside of the United States excluding those with U.S. parents. Groups are defined by country-of-origin information. The ethnicities shown here are based upon answers to the first ancestry code on the census. The sample does not exclude students, the self-employed, and persons with zero earnings

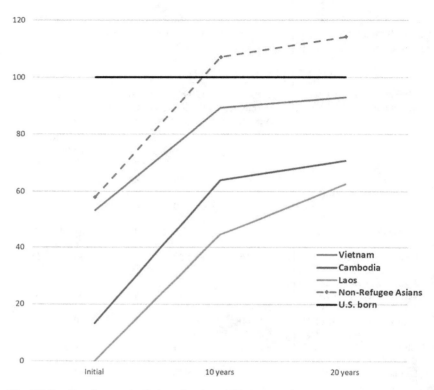

Fig. 19.1 Earnings assimilation, for the 1975–1980 cohort, relative to U.S. born, measured with median earnings, for Indochinese refugee groups and the developing country Asian immigrants. (Source: Authors' creation based on 1980, 1990, 2000 census PUMS data)

assimilation is greatest in the first ten years. Yet, it still continues in the tenth to twentieth year, particularly for refugee men from Laos.

INVESTMENT IN U.S. HUMAN CAPITAL

English Proficiency

With 45% reporting poor or nonexistent English during their initial U.S. years (Table 19.5), the 1975–1980 cohort of Indochinese refugee men resembles some of the developing-country Asian immigrant groups of Part II: 46% of Korean and 42% of Chinese immigrant men in the

Table 19.5 English proficiency of the 1975–1980 cohort of immigrant men during the first five years and ten years later

Place of birth in South East Asian:	Speaks English poorly or not at all				Speaks English very well			
	1980, ages 25–54	1990, ages 35–64	Change in percent	Relative to Non-South East Asian	1980, ages 25–54	1990, ages 35–64	Change in percent	Relative to Non-South East Asian
Indochina	45.2	28.1	−18.1	2.55	16.6	26.3	9.7	0.80
Vietnam	38.7	23.2	−15.5	2.18	18.8	28.3	9.5	0.78
Cambodia	66.7	37.7	−29.0	4.08	7.6	24.3	16.7	1.37
Laos	68.6	47.4	−21.2	2.99	9.1	17.4	8.3	0.68
Ethnicity (born in Indochina):								
Vietnamese	34.1	20.6	−13.5	1.90	20.3	30.9	10.6	0.87
Cambodian	67.1	36.3	−30.8	4.34	7.9	27.1	19.2	1.57
Chinese	63.1	35.6	−27.5	3.87	10.8	17.9	7.1	0.58
Laotian	71.0	47.3	−23.7	3.34	8.4	16.3	7.9	0.65
Hmong	73.6	52.5	−21.1	2.97	5.7	11.9	6.2	0.51
Developing country Asian immigrants								
China, India, Korea, Philippines	23.7	16.6	−7.1	1.00	39.3	51.5	12.2	1.00

Source: Authors' estimates based on a 6% 1980 file created by combining and reweighting the 5% and 1% files of the 1980 Census of Population and a 6% microdata 1990 sample created by combining and reweighting the 1990 Census of Population Public Use 5% and 1% samples. Appendix A provides information on sample sizes for entry cohorts at entry and ten years later

Notes: Native born are defined as persons born in the U.S.; foreign born are defined as persons born outside of the United States excluding those with U.S. parents. Groups are defined by country-of-origin information. The ethnicities shown here are based upon answers to the first ancestry code on the census. The sample does not exclude students, the self-employed, and persons with zero earnings

1975–1980 cohort spoke English poorly (Chap. 9).[1,2] Dividing by ethnicity reveals a range of initial proficiency levels. About a third of the Vietnamese spoke English poorly compared with over two-thirds in the other ethnic groups. Nearly three-quarters of the Hmong spoke English poorly.

[1] The rate for all Asian immigrants reporting "poor" or "not at all" was a much lower 27.5%, driven by very high proficiency rates from former colonies of English-speaking countries—principally India and the Philippines.
[2] The 16.2% of Indochinese refugee men in the 1975–1980 cohort who reported that they spoke English "very well" also approximates the rates of these groups.

Regardless of starting point, the English proficiency of all Indochinese groups grew with time in the United States. Most notable is the dwindling fraction of those who spoke English poorly or not at all. After 10 to 15 years in the United States, the percentage of poor English speakers among the Indochinese approached the 26% rate for the 1975–1980 cohort of immigrants from all countries. Even for the Indochinese ethnic groups with the greatest initial proportion of poor English speakers (the Hmong and the Laotians), the poorly proficient decrease by more than 20 percentage points.[3]

Despite extremely low education levels, the non-Vietnamese refugees' English improvement exceeds that of the 1975–1980 cohort of immigrants from non-English-speaking Western Europe (Chap. 9). When measured by the percentage who speak English poorly or not at all, the English improvement among the refugee groups is two to four times that of the developing country Asian immigrants. When measured by the percentage who speak English well, all refugee groups, other than the Cambodians, show lower improvement than the developing country Asian immigrants.

Investment in Education

In regression analyses of immigrant earnings profiles, economists generally hold constant the schooling level of adult immigrants even though the education of immigrants, particularly those with low skill transferability, is far from fixed. In 1980, 38% of 25- to 54-year-old Indochinese refugee men, who entered the United States between 1975 and 1980, report more than 12 years of schooling. In 1990, for the same cohort, aged ten years, 53% report more than a high school diploma.

Immigrant emigration and changes in how individuals identify themselves make it difficult to measure human capital investment with changes across decennial censuses in English proficiency, occupational change, and educational attainment. For instance, the 1975–1980 Hmong population doubles between censuses. Perhaps with time in the United States, many Hmong who reported "Laotian" in 1980 acquire a distinct ethnic identity, or at least the ability to express it on a census form. Because it measures human capital investment as it occurs, school attendance provides a

---

[3] The correct measure of change in English proficiency (absolute or percentage) depends on whether one thinks that the difficulty in learning a new language changes with the initial proficiency level or is independent of the initial proficiency level.

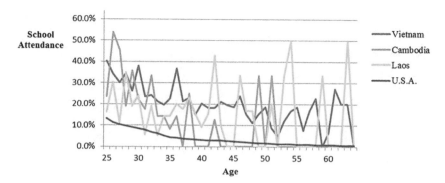

Fig. 19.2 Percentage attending school by single years of age, 1980. (Source: Authors' creation based on 1980 census PUMS data)

more certain source of information on one form of human capital investment.

Among those entering the United States between 1975 and 1980, almost all of the refugee groups attended school in 1980 at rates that equaled or exceeded the attendance rates of the developing-country Asian immigrants, whose rates exceeded those of U.S. natives. Although school attendance decreases with age for U.S. natives, the higher school attendance of the Indochinese persists regardless of age (Fig. 19.2). It seems unlikely that this educational investment reflects English language training per se since the census question asks only about courses in degree or high school diploma programs.

The greater propensity to invest in schooling persists with time in the United States. All of the Indochinese ethnic groups have higher in-school rates 10–15 years after entry than the developing country Asian immigrants, whose own school attendance is more than double that of the West Europeans (Table 19.6).[4] This is particularly true of the Hmong, of whom a fifth of the 1975–1980 cohort attended school in 1990.

---

[4] In 1990, we looked here only at those aged 35–64 to be comparable with the 25–54 cohort examined in 1980.

Table 19.6 School attendance of the 1975–1980 cohort during first five years and ten years later: immigrant men (percent)

Place of Birth in South East Asia:	1980, ages 25–54	1990, ages 35–64
Indochina		
Vietnam	25.6	10.3
Cambodia	21.2	11.3
Laos	18.1	15.8
Ethnicity (born in Indochina):		
Vietnamese	26.1	10.5
Cambodian	19.3	13.1
Chinese	19.9	10.2
Laotian	15.9	11.7
Hmong	24.5	20.5
Developing Country Asian Immigrants		
China, India, Korea, Philippines	19.1	6.3

Source: Authors' estimates based on a 6% 1980 file created by combining and reweighting the 5% and 1% files of the 1980 Census of Population and a 6% microdata 1990 sample created by combining and reweighting the 1990 Census of Population Public Use 5% and 1% samples. Appendix A provides information on sample sizes for entry cohorts at entry and ten years later

Notes: Native born are defined as persons born in the U.S.; foreign born are defined as persons born outside of the United States excluding those with U.S. parents. Groups are defined by country-of-origin information. The ethnicities shown here are based upon answers to the first ancestry code on the census. The sample does not exclude students, the self-employed, and persons with zero earnings

Concluding Remarks

This chapter describes a diversity of experiences for the Indochinese refugee groups. Among the more educated, who were predominantly Vietnamese, the Indochinese began their U.S. lives with higher earnings and lower earnings growth than did the 1975–1980 cohort of developing-country Asian immigrants. The pattern suggests that for this first cohort, the more educated Indochinese had higher skill transferability to the United States than did the 1975–1980 cohort of developing-country Asian immigrants. Reimers (2005, p. 282) notes:

> Among the first Vietnamese to leave were an elite, many with connections to American firms or governmental agencies, including service with the military.

Examining the Indochinese by ethnicity reveals that the ethnic Vietnamese had higher initial education levels and English proficiency than the ethnic Chinese: 23% of ethnic Vietnamese had less than twelve years of schooling versus 52% of the Chinese; 34% of the Vietnamese speak English poorly or not at all versus 63% for the Chinese (Tables 22.1 and 19.5). Measured at the mean, the Chinese began their U.S. economic journeys with lower initial earnings than the Vietnamese. Yet, ten years later, the Chinese earn 93% of U.S. natives' earnings versus 86% for the Vietnamese. Twenty years later, in results not presented, their mean earnings are 103% of U.S. natives' earnings versus 89% for the Vietnamese.

Following the 1975–1980 cohort of immigrants, Cortes (2004) compared refugees with economic migrants. She concludes that the refugees' implied greater permanence explains their higher human-capital investment and earnings growth. Yet, our estimated emigration rates (Chap. 18) suggest that the Indochinese refugee groups were no more permanent than Asian immigrants from economically developing countries, who were highly permanent. What distinguishes the Indochinese refugees from their developing-country non-refugee counterparts is not the degree of their permanence, but its nature. For non-refugee Asian immigrants, the decision to stay in the U.S. was voluntary; for refugees, it was forced.

Our intergroup comparisons suggest the following generalizations:

- Permanent immigrants with low skill transferability will have relatively high propensities to invest in host-country human capital regardless of whether their permanence is voluntary or forced.
- Despite their permanence, refugees with highly transferable skills will have a lower propensity to invest in host-country human capital than will permanent immigrants with low skill transferability.

References

Cortes, K. E. (2004) "Are Refugees Different from Economic Immigrants? Some Empirical Evidence on the Heterogeneity of Immigrant Groups in the United States," *Review of Economics and Statistics*, 2004, 86 (2), 465–480

Reimers David M. 2005. *Other Immigrants: The Global Origins of the American People*, NY: NYU Press

CHAPTER 20

Married Refugee Women from South East Asia: The 1975–1980 Cohort

The Vietnamese married women who came to the United States in the years 1975 to 1980 resemble the 1975–1980 cohort of Western European married women on two human capital counts (Table 20.1). Both groups have about ten years of schooling and 56% of the Vietnamese versus 52% of the West Europeans speak English poorly. In stark contrast, Cambodian and Laotian wives average less than six years of schooling and 85% speak English poorly.

On average, Vietnamese, Cambodian, and Laotian families have more than three children at home; West Europeans have fewer than two. The South East Asian groups are more likely to have young children at home; almost a quarter of the Cambodians and Laotians have a baby at home. The husbands of South East Asian refugee women have entry-level earnings far below the average for West European husbands.[1] Vietnamese husbands—like the developing-country Asian groups featured in Part II—earn less than half the West European benchmark; Cambodian and Laotian husbands earn a quarter or less of the West European earnings. Women in all of the refugee groups have higher disability rates than the West Europeans.

---

[1] There are additional signs that Cambodian and Laotian families are at a particular disadvantage upon entry into the U.S. Both groups have high unemployment rates among husbands and almost no interest earnings, an indication of virtually no assets.

© The Author(s), under exclusive license to Springer Nature Switzerland AG 2020
H. Duleep et al., *Human Capital Investment*,
https://doi.org/10.1007/978-3-030-47083-8_20

Table 20.1 Mean characteristics in 1980 of married immigrant women from South East Asia: 25–54 years old, immigrated between 1975 and 1980

	South East Asian refugee groups			Western Europe
	Vietnam	Cambodia	Laos	
Human capital				
Experience	18.77	21.34	21.07	19.51
Education	10.07	5.57	5.66	9.83
Percentage who speak English not well or not at all	0.56	0.86	0.85	0.52
Disabled	0.04	0.05	0.04	0.02
Children				
Number	3.08	3.15	3.77	1.82
Age distribution				
Child under 1	0.12	0.25	0.23	0.08
Child 1 to 5	0.53	0.64	0.73	0.37
Child 6 to 11	0.60	0.64	0.70	0.42
Child 12–17	0.40	0.34	0.43	0.31
Husband				
Assets	$111	$18	$29	$845
Earnings	$14,161	$7631	$5978	$30,577
Unemployment	0.05	0.08	0.08	0.04
Yp-Yc	$19,062	$15,484	$18,794	$1990

Source: Authors' estimates based on a 6% 1980 file created by combining and reweighting the 5% and 1% files of the 1980 Census of Population. Appendix A provides information on sample sizes by entry cohort

Notes: Foreign born are defined as persons born outside of the United States excluding those with U.S. parents. Groups are defined by country-of-origin information. The sample does not exclude students, the self-employed, and persons with zero earnings. The sample of married immigrant women is restricted to women who are married to foreign-born men. War brides are not included

When we look at the gap between the potential earnings (Y_p) of immigrant husbands and their current earnings (Y_c), the current earnings of West European husbands are close to their earnings potential whereas South East Asian husbands earn far below their potential earnings (last row of Table 20.1).[2] In Part III, we learned that immigrant married women were more likely to work and to work more hours in all groups in which the husbands had low initial earnings relative to their potential

---

[2] Following the procedure described in Chap. 13, we measure the husband's potential earnings using non-parametric estimates of the earnings of U.S.-born men in 1980 who are the same age and living in the same areas with equal levels of schooling.

earnings. This finding, which emerged not controlling for factors that affect the reservation and market wages of married immigrant women, prevailed controlling for a full panoply of relevant variables. It is the positive relationship between the adjusted earnings gap for men and the propensity to work of immigrant women that creates the dramatic change in relative economic status when a family perspective is adopted (Table 11.1).

Among married immigrant women in the United States for five or fewer years in 1980, and not adjusting for any variables, Vietnamese women are more likely to work than the West Europeans: 53% versus 47% (Table 20.2). The propensity to work of Cambodian and Laotian women, however, lies far below the West European mark.

Cambodian and Laotian women have low levels of schooling, low levels of English proficiency, and large families, all factors that decrease the propensity of women to work. What happens when we control for these variables?

Modeling the Propensity of South East Asian Women to Work

To probe the determinants of South East Asian immigrant women working, we estimated a pooled regression model that controls for general skills, family structure (number and ages of children), education, potential labor market experience, geographic location (region of residence, urban location), and disability status.[3] The logit coefficients from this model for the group variables and their associated marginal effects[4] are shown in the third and fourth columns of Table 20.2. To this basic model, English proficiency is added (last two columns).[5]

---

[3] A more general specification would be to estimate separate regression models by immigrant group and then evaluate each model at a common level of general human capital and family structure. This is more appropriate if the returns to general human capital or the effects of family structure on labor supply are different across immigrant groups. We chose to estimate a pooled regression for two reasons. First, our sample sizes for immigrants from Cambodia and Laos are small; regression models for these groups would be imprecise. Second, separate regressions for West Europeans and Vietnamese wives showed that while some point estimates are quantitatively different, most are quantitatively similar and almost all are qualitatively similar.

[4] The marginal effects are evaluated at the average probability of working.

[5] We estimated a model that also included husband's earnings, assets, and unemployment status. These results are not discussed because they differ so little from the more basic model.

Table 20.2 The propensity to work of married immigrant women from South East Asia during their first five years in the U.S. and ten years later: Cohort who immigrated between 1975 and 1980 (asymptotic t-statistics in parentheses)

	Unadjusted group mean	Unadjusted group effect	Adjusted for general skills and characteristics		Adjusted for English proficiency	
			logit coef.	$\partial p/\partial x$	logit coef.	$\partial p/\partial x$
Propensity to work in 1979, the first 5 years						
Vietnamese	0.528	0.06	0.8336 (7.13)	0.21	0.8504 (7.14)	0.21
Cambodian	0.303	-0.17	-0.0883 (0.32)	-0.02	-0.0166 (0.058)	-0.00
Laotian	0.221	-0.25	-0.5168 (2.64)	-0.13	-0.4328 (2.18)	-0.11
W. European	0.468		–		–	
Propensity to work in 1989, 10 years later						
Vietnamese	0.733	0.11	1.1271 (7.08)	0.25	1.1346 (6.90)	0.25
Cambodian	0.644	0.02	0.8609 (2.85)	0.19	0.8853 (2.88)	0.19
Laotian	0.552	-0.07	0.8929 (3.80)	0.20	0.9262 (3.88)	0.20
West European	0.623		–		–	

Source: Authors' estimates based on a 6% 1980 file created by combining and reweighting the 5% and 1% files of the 1980 Census of Population and a 6% microdata 1990 sample created by combining and reweighting the 1990 Census of Population Public Use 5% and 1% samples. Appendix A provides information on sample sizes for entry cohorts at entry and ten years later

Notes: Foreign born are defined as persons born outside of the United States excluding those with U.S. parents. Groups are defined by country-of-origin information. The sample does not exclude students, the self-employed, and persons with zero earnings. The sample of married immigrant women is restricted to women who are married to foreign-born men. War brides are not included. West Europeans, excluding English-speaking countries, form the reference group. Labor force participation is defined as positive earnings and positive hours and weeks worked during the year preceding the census. All regressions include the following set of variables: child status, potential labor market experience, education, region of residence and an indicator for urban residence

The study population includes immigrant wives, ages 25–54 in 1980, 0–5 years after entering the United States (top panel), and 10 years later, in 1990, after these women (ages 35–64) have been in the United States for 10–15 years (bottom panel). Given concerns about the high emigration rate of English-speaking West Europeans, the benchmark group is non-English-speaking West Europeans. For each of these regressions, the estimated effects are relative to this group whom we will simply refer to as West European.

The top panel shows that when we adjust for general skills and characteristics, the greater propensity to work of the Vietnamese relative to the West European benchmark increases. The increase likely reflects adjusting for the greater number of children in Vietnamese homes since the English proficiency and schooling levels of the Vietnamese and West Europeans are similar. Adjusting for general skills and characteristics and English proficiency halves the labor force participation gap between Laotians and the West Europeans, while eliminating it for Cambodians.

With time in the United States, labor force participation increases for married women in all immigrant groups (bottom half of Table 20.2). Not adjusting for any variables, Vietnamese wives continue to be more likely to work than their West European counterparts; adjusting for general skills and characteristics, the Vietnamese effect increases.

Of particular interest is the dramatic change in the propensity to work for Cambodian and Laotian wives. After 10–15 years in America—and not adjusting for any variables—Cambodian women are more likely to work than their West European counterparts. Married Laotian women are still seven percentage points less likely to work than West European women, when no variables are controlled for. Adjusting for general human capital and family structure, Cambodian and Laotian wives have labor force participation rates 19 and 20 percentage points higher than their West European counterparts.[6]

HOURS OF WORK: THE FIRST FIVE YEARS AND TEN YEARS LATER

Table 20.3 presents the effect of being in each source-country group on hours worked of labor force participants. To be in the sample, a woman must have positive earnings. The results, as such, do not reflect variations in labor force participation. The top half shows the hours worked in 1980, up to five years after entering the United States; the bottom half shows this ten years later.

---

[6] We also investigated an alternative definition of labor force participation, working for pay or working at unpaid labor as migrant wives may work in businesses with husbands. The results are nearly identical for all groups.

Conforming in spirit to the labor force participation results that do not adjust for any variables—or adjusting for all relevant variables—Vietnamese wives work more hours during their initial U.S. years than the West Europeans. Ten years later, Vietnamese wives continue to work many more hours than their West European counterparts.

Not adjusting for general characteristics and skill differences, Cambodian and particularly Laotian wives work fewer hours during their first five U.S. years than the West Europeans. Adjusting for general characteristics and skills increases the hours worked of recently arrived Cambodian and Laotian married women so that their hours worked are not statistically significantly different from the hours worked of West European married

Table 20.3 Hours worked for working married immigrant women from South East Asia during their first five years in the U.S. and ten years later: Cohort who immigrated between 1975 and 1980 (t-statistics in parentheses)

	Unadjusted mean	Unadjusted South East Asian effect	Adjusted for general skills and characteristics	Adjusted for English proficiency
Hours worked in 1979, the first 5 years				
Vietnamese	1608.4	117.6	241.9 (3.57)	236.2 (3.43)
Cambodian	1460.6	-30.2	82.0 (0.47)	85.1 (0.49)
Laotian	1362.8	-128.0	-8.8 (0.07)	-2.1 (0.02)
West European	1490.8			
Hours worked in 1989, ten years later				
Vietnamese	1935.3	213.6	299.3 (4.54)	296.7 (4.40)
Cambodian	1957.8	236.1	307.7 (2.45)	299.8 (2.38)
Laotian	1762.5	40.8	136.5 (1.37)	142.0 (1.41)
West European	1721.7			

Source: Authors' estimates based on a 6% 1980 file created by combining and reweighting the 5% and 1% files of the 1980 Census of Population and a 6% microdata 1990 sample created by combining and reweighting the 1990 Census of Population Public Use 5% and 1% samples. Appendix A provides information on sample sizes for entry cohorts at entry and ten years later

Notes: Foreign born are defined as persons born outside of the United States excluding those with U.S. parents. Groups are defined by country-of-origin information. The sample of married immigrant women is restricted to women who are married to foreign-born men. War brides are not included. In order to be in the sample, a woman had to report positive hours worked, positive weeks worked, and positive earnings. All regressions include the following set of variables: child status, potential labor market experience, education, region of residence and an indicator for urban residence. West Europeans, excluding English-speaking countries, form the reference group in all of the pooled regressions

women. Ten years later—*and not adjusting for any variables*—Cambodian and Laotian wives work more hours than their West European counterparts.

The Earnings of South East Asian Married Women: The First Five Years and Ten Years Later

Table 20.4 compares the earnings at entry and ten years later of South East Asian married immigrant women with those of West European women. During their initial U.S. years, and not adjusting for any factors, Vietnamese wives earn substantially more than the West Europeans. The Vietnamese earnings' advantage over the West Europeans increases when we adjust for general skills and characteristics. Since the sample does not include women with zero earnings, the increase for Vietnamese women is not due to increases in labor force participation.

Married women from Cambodia and Laos initially earn substantially less than West European wives do during the initial years. Adjusting for skills and characteristics and English proficiency, eliminates this difference. Yet, unlike the Vietnamese and the developing-country Asian immigrants, Cambodian and Laotian women do not earn more than statistically similar West European wives in their initial U.S. years.

With time in the United States, the earnings of all immigrant wives increase substantially. At the 10–15-year mark, and not adjusting for any characteristics, the 1989 earnings of married women from Vietnam and Cambodia far exceed the average earnings of the 1975–1980 cohort of West European wives, who also experienced substantial earnings gains since 1980. The 1989 unadjusted earnings of Laotian wives, though triple their earnings a decade earlier, are still below the West European benchmark. Yet, once we control for general skills and characteristics, Laotian wives outearn their West European statistical twins by $3352.

The Wages of Working South East Asian Married Immigrant Women: The First Five Years and Ten Years Later

Immigrant women who are married to men with high returns to investing in U.S.-specific skills might be less likely to undertake human capital investments (or to delay such investments) to take jobs that pay more

Table 20.4 The earnings of married immigrant women from South East Asia at entry and ten years later (in 1990 dollars): Cohort who immigrated between 1975 and 1980 (t-statistics in parentheses)

	Unadjusted mean	Unadjusted South East Asian effect	Adjusted for general skills and characteristics	Adjusted for English proficiency
Earnings in 1979, the first five years				
Vietnamese	6126.4	1180	2973 (7.33)	3074 (7.54)
Cambodian	2188.7	-2759	-478 (0.51)	-122 (0.12)
Laotian	2024.6	-2923	-418 (0.67)	-2 (0.00)
West European	4947.9			
Earnings in 1989, ten years later				
Vietnamese	13,489.8	4548	5905 (6.81)	6017 (6.81)
Cambodian	10,507.6	1566	4707 (2.84)	4844 (2.92)
Laotian	6879.6	-2063	3352 (2.58)	3641 (2.79)
West European	8941.7			

Source: Authors' estimates based on a 6% 1980 file created by combining and reweighting the 5% and 1% files of the 1980 Census of Population and a 6% microdata 1990 sample created by combining and reweighting the 1990 Census of Population Public Use 5% and 1% samples. Appendix A provides information on sample sizes for entry cohorts at entry and ten years later

Notes: Foreign born are defined as persons born outside of the United States excluding those with U.S. parents. Groups are defined by country-of-origin information. The sample of married immigrant women is restricted to women who are married to foreign-born men. War brides are not included. In order to be in the sample, a woman had to report positive hours worked, positive weeks worked, and positive earnings. All regressions include the following set of variables: child status, potential labor market experience, education, region of residence and an indicator for urban residence. West Europeans, excluding English-speaking countries, form the reference group in all of the pooled regressions.

during the period in which their husbands' investment in U.S.-specific skills is greatest. Their initial wages would be higher than would otherwise be the case and their wage growth lower.

Figure 20.1 presents the wages of South East Asian working wives in their first five years in the United States and ten years later. Their initial wages are below those of West Europeans. Yet, over time, their wages grow faster than those of their West European counterparts. In ten years,

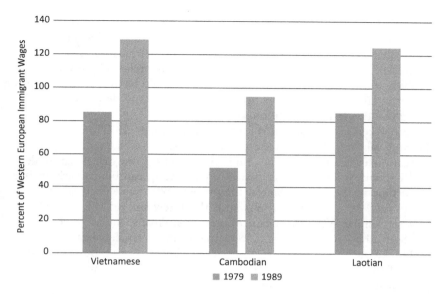

Fig. 20.1 The wages of Indochinese refugee married women as a percentage of Western European wages (1975–1980 entry cohort, married women) 25–54 in 1980, 35–64 in 1990. (Source: Authors' creation based on 1980 and 1990 census PUMS data)

Vietnamese and Laotians gain a wage advantage over the West Europeans, and the Cambodians almost close the wage gap that they initially faced. These unadjusted results are all the more remarkable given the low education levels of the Cambodians and Laotians.[7]

Concluding Remarks

Despite their overwhelming refugee status, Vietnamese women resemble the Asian developing-country groups studied in Part III, of whom few are refugees. Evident in their first U.S. years, they show a large unexplained propensity to work. Their high labor force participation during their initial years in the United States counterbalances the low adjusted earnings of their husbands. The Family Investment Model predicts that the greater

---

[7] The same pattern of relatively low initial wages followed by coming close to or overtaking the West European benchmark persists in multivariate analyses.

the adjusted earnings gap of men, the more likely that married women will work. Given that the adjusted earnings gap is greatest for Vietnamese men, this FIM prediction holds. Among Vietnamese men, the more educated had relatively high entry earnings and relatively low earnings growth. As an avenue for further research, it would be interesting to learn how the labor force participation of Vietnamese married women varies when you divide by education level.

Other key predictions of the Family Investment Model do not hold. During their initial years in the United States, and adjusting for their low education levels and family structure, the propensity to work of Cambodian and Laotian women is lower than the West European benchmark group. With time in the United States, the propensity to work increases for the South East Asian refugee groups both absolutely and relative to their statistically similar West European counterparts. Finally, the wage growth of the South East Asian groups exceeds that of the West Europeans.

Unlike the Vietnamese and Asian developing-country groups, the earnings of the Cambodians and Laotians do not counterbalance the initial low earnings of their husbands. This picture changes with time for the Cambodians.

CHAPTER 21

Refugee Entrants from South East Asia, a Decade After the War: The 1985–1990 Cohort

Although the Vietnam War ended in 1975, the refugee flow from it—and the associated wars in Laos and Cambodia—continued for over two decades. Many leaving well after 1975 were "boat people," who left Vietnam in small vessels, arriving not in the United States or another permanent host country, but in a Southeast Asian refugee camp where they would remain for years. Only in 1996 were most refugee camps closed.

First-asylum nations in Southeast Asia, the U.N. High Commission on Refugees, and the major resettlement countries agreed to a camp-closing schedule—the Comprehensive Plan of Action (CPA). Under the CPA, the U.N. High Commission on Refugees screened potential immigrants for economic versus political motivations, with the major resettlement countries agreeing to take the latter. The CPA led to large increases in Vietnamese admissions to the United States, and voluntary or forced repatriation of those deemed economic migrants.[1] After April 1996, Vietnamese refugees were required to return to Vietnam before applying for political asylum. According to the 1990 census, Vietnamese made up 60% of the

---

[1] There appears to be no good count available of voluntary versus forced repatriations. Even if such a count were available, allegations of various forms of coercion to volunteer make the distinction problematic. For example, the May 1, 1995 Wall Street Journal reported that in Hong Kong, prisoners were released directly into the camp population to threaten residents who did not volunteer. As of April 1996, the UNHCR reported repatriating 79,000 Vietnamese. Repatriates received a reintegration grant and pocket money worth approximately $290.

© The Author(s), under exclusive license to Springer Nature Switzerland AG 2020
H. Duleep et al., *Human Capital Investment*,
https://doi.org/10.1007/978-3-030-47083-8_21

Indochinese total. After 1990, this proportion grew, reflecting the largely Vietnamese presence in the camps affected by the CPA.

Like the 1975–1980 cohort, the more recent migrants were mostly refugees. Nearly 20 years after the Vietnam War ended, only 3423 of the 41,345 Vietnamese given permanent U.S. visas in 1994 were admitted via a family admissions category; 113 entered via an employment category. Thus, our comments in Chap. 18 about potential differences between refugees and immigrants pertain to the newest entrants, with one critical caveat: to the likely detriment of labor market skills and health,[2] many entering the United States under the CPA had spent up to ten years in refugee camps throughout East Asia.

THE ENTRY CHARACTERISTICS OF THE 1975–1980 AND 1985–1990 COHORTS

In comparing the education levels of the 1975–1980 and 1985–1990 cohorts, the 1980 census measures schooling achievement by the number of years completed; the 1990 census measures it by a harder standard—degrees completed. With this in mind, the education levels of the developing-country Asian immigrants (the groups studied in Parts II–IV) changed very little. In 1980, 30% reported 12 years or less of schooling; in 1990, 28% reported completing a high school degree. In 1980, 30% reported completing 16 or 17 years; in 1990, 27% reported having a Bachelor of Arts or Bachelor of Sciences degree (Table 21.1). Concomitantly, the education levels of West European immigrants increased dramatically. English proficiency for the developing-country Asian immigrants also changed little from 1980 to 1990, while improving for West European immigrants (Table 21.2).

In contrast, the entry-level schooling levels of Indochinese refugee men plummeted: 35% of the 1975–1980 cohort had less than 12 years of schooling versus 53% of the 1985–1990 cohort (Table 21.1). Dividing by country and ethnic group reveals little or no change for most groups. The steep decline in Indochinese educational attainment reflects the Vietnamese experience for whom the percentage with less than 12 years of schooling doubled from 1980 to 1990.

---

[2] Reimers (2005) provides an in-depth exploration of the health issues, including mental health issues, accompanying the Indochinese refugee experience.

Table 21.1 Initial education levels of the 1975–1980 and 1985–1990 entry cohorts: Immigrant men, 25–54 years old (percent)

	Less than 12 years	12	13–15	16–17	18 or more years	Less than high school	High school degree	Some College	BA/BS degree	Graduate degree
Place of birth in South East Asia										
Vietnam	27.4	30.3	27.4	8.8	6.0	48.1	20.6	21.9	6.2	3.3
Cambodia	58.5	16.1	17.1	5.9	2.4	64.4	10.9	17.0	4.9	2.7
Laos	61.1	20.3	10.7	5.8	2.1	60.0	15.0	18.0	4.4	2.6
Ethnicity (born in Indochina)										
Vietnamese	22.9	30.0	30.9	9.4	6.8	46.5	21.2	21.8	6.4	4.1
Cambodian	59.3	17.9	14.5	5.5	2.8	63.7	11.4	18.9	3.6	2.5
Chinese	52.1	30.3	10.9	4.7	2.0	58.0	17.7	17.9	4.8	1.6
Laotian	62.5	21.0	9.8	5.2	1.5	59.1	17.5	16.5	3.9	3.1
Hmong	62.3	18.9	15.1	3.8	0.0	65.0	8.2	21.0	4.0	1.7
Non-South East Asian comparison groups										
China, India, Korea, Philippines	14.4	15.9	17.5	30.4	21.8	14.2	14.0	17.3	27.1	27.3
Western Europe	34.5	19.3	13.1	16.2	16.9	18.4	15.9	18.3	16.4	31.0

Source: Authors' estimates based on a 6% 1980 file created by combining and reweighting the 5% and 1% files of the 1980 Census of Population and a 6% microdata 1990 sample created by combining and reweighting the 1990 Census of Population Public Use 5% and 1% samples. Appendix A provides information on sample sizes by entry cohort.

English proficiency also deteriorated for the Indochinese refugees. The percentage with poor or nonexistent English increased from 46% for the 1975–1980 cohort to 61% for the 1985–1990 cohort (Table 21.2). Again, the deterioration reflects the Vietnamese experience; English proficiency of the other groups barely changed or improved.

From 1980 to 1990, the disability rates of Indochinese refugees more than doubled (Table 21.3). Though the Vietnamese still have the lowest rate, it is double the 1980 rate. For all the other Indochinese groups, disability rates increased substantially. Especially notable is the fourfold increase for Cambodians. The disability surge likely reflects that many refugees in the 1985–1990 cohort had spent years in refugee camps. The high Cambodian rate may stem from the internecine war in Cambodia following the 1979 Vietnamese invasion to dislodge the Khmer Rouge

Table 21.2 Initial English proficiency of the 1975–1980 and 1985–1990 entry cohorts: immigrant men, 25–64 years old (percent)

	1980		1990	
	Speaks English poorly or not at all	Speaks English very well	Speaks English poorly or not at all	Speaks English very well
Place of birth in South East Asia				
Indochina	46.2	16.2	61.1	13.5
Vietnam	38.7	18.8	58.3	12.7
Cambodia	66.7	7.6	62.7	19.0
Laos	68.6	9.1	67.6	12.3
Ethnicity (born in Indochina)				
Vietnamese	34.1	20.3	56.6	13.3
Cambodian	67.1	7.9	64.5	17.9
Chinese	63.1	10.8	66.8	9.5
Laotian	71.0	8.4	67.7	12.0
Hmong	73.6	5.7	71.4	9.6
Non-South East Asian comparison groups				
China, India, Korea, Philippines	26.0	37.8	27.8	38.4
Western Europe	30.7	41.4	26.6	49.4

Source: Authors' estimates based on a 6% 1980 file created by combining and reweighting the 5% and 1% files of the 1980 Census of Population and a 6% microdata 1990 sample created by combining and reweighting the 1990 Census of Population Public Use 5% and 1% samples. Appendix A provides information on sample sizes by entry cohort

government. Public assistance, which correlates with high disability rates and low education, is highest for the Hmong in both 1980 and 1990 (Table 21.4).

ENTRY EARNINGS AND EARNINGS GROWTH

From 1980 to 1990, the entry earnings of the developing-country Asian immigrants, ages 25–54, decreased some. The entry earnings of the Indochinese fell markedly, from 38% of U.S. natives' earnings to just 14% of the U.S.-born benchmark (Table 21.5). This earnings decline primarily reflects a large decrease in the entry earnings of the Vietnamese and some decline in the entry earnings of Cambodians and Laotians. The drop in Vietnamese earnings can be attributed to the substantial drop in their entry-level schooling and English proficiency whereas the schooling levels

Table 21.3 Initial disability rates of the 1975–1980 and 1985–1990 entry cohorts: Immigrant men, 25–64 years old (percent)

	1975–80	1985–90
Place of birth in South East Asia		
French Indochina	5.1	11.8
Vietnam	4.5	8.6
Cambodia	5.6	22.1
Laos	7.8	14.3
Ethnicity (born in Indochina):		
Vietnamese	3.9	9.0
Cambodian	5.7	21.9
Chinese	6.5	10.4
Laotian	7.4	14.1
Hmong	11.3	15.4
Non-South East Asian comparison groups		
China, India, Korea, Philippines	3.8	2.9
Western Europe	2.1	3.6

Source: Authors' estimates based on a 6% 1980 file created by combining and reweighting the 5% and 1% files of the 1980 Census of Population and a 6% microdata 1990 sample created by combining and reweighting the 1990 Census of Population Public Use 5% and 1% samples. Appendix A provides information on sample sizes by entry cohort

and English proficiency for entering Cambodians and Laotians stayed fairly constant.

Measuring earnings by the median (Table 21.5), and within either the 1975–1980 or 1985–1990 cohort, we find that across the Indochinese country-of-origin and ethnic groups, there is an inverse relationship between entry earnings and earnings growth—the lower the initial earnings, the higher the earnings growth—even with no adjustment for initial human capital levels. When measured at the mean (Table 21.6), exceptions to the "unadjusted" inverse relationship appear. Relative to other Asian refugee or developing country groups, Laotians, defined by country of origin or ethnicity, have lower initial earnings *and* lower earnings growth, as do the Hmong in the 1975–1980 cohort. When we follow the same groups across cohorts, declines in entry earnings are generally associated with increases in earnings growth. An exception are Cambodian refugees. Whether defined by country of origin or ethnicity, both their initial earnings and their earnings growth fell from the 1975–1980 cohort to the 1985–1990 cohort.

Table 21.4 Receipt of public assistance income during the first five years by the 1975–1980 and 1985–1989 entry cohorts: Immigrant men, 25–64 years old (percent)

	1975–80 entry cohort	1985–90 entry cohort
	1979	1989
Place of birth in South East Asia		
Indochina	17.8%	29.2%
Vietnam	16.3	21.4
Cambodia	21.7	44.7
Laos	24.0	41.4
Ethnicity (born in Indochina)		
Vietnamese	14.5	22.1
Cambodian	22.1	49.1
Chinese	26.2	21.3
Laotian	23.6	36.1
Hmong	32.1	57.8
Non-South East Asian comparison groups		
China, India, Korea, Philippines	1.1	0.9
Western Europe	1.1	7.6

Source: Authors' estimates based on a 6% 1980 file created by combining and reweighting the 5% and 1% files of the 1980 Census of Population and a 6% microdata 1990 sample created by combining and reweighting the 1990 Census of Population Public Use 5% and 1% samples. Appendix A provides information on sample sizes by entry cohort

Examining economic assimilation by ethnicity, a stunning statistic previously noted in the 1975–1980 cohort (Chap. 19) re-emerges in the 1985–1990 cohort. Even though the ethnic Vietnamese have both higher education and higher English proficiency at entry than the ethnic Chinese, ten years later, men of Vietnamese ethnicity earn 59% of the mean earnings of U.S. natives whereas those of Chinese ethnicity earn 96% of U.S.-born earnings (Table 21.6).

HUMAN CAPITAL INVESTMENT PATTERNS

Despite their extremely low education levels and high disability rates, school attendance among the 1985–1990 cohort of Indochinese refugee men, for all groups, favorably compares with those of the developing country Asian immigrants. Nearly half of the Hmong report attending school (Table 21.7). For Vietnamese men, the educational level of

Table 21.5 Median entry earnings, earnings ten years later, and ten-year growth rates for the 1975–1980 and 1985–1990 entry cohorts; Earnings at entry and ten years later are expressed as a percentage of U.S. native-born median earnings

	1975–1980 cohort			1985–1990 cohort		
	Ages 25–54	Ages 35–64		Ages 25–54	Ages 35–64	
	1979 Earnings	1989 Earnings	10-Year growth rate	1989 Earnings	1999 Earnings	10-Year growth rate
Place of birth						
Indochina	38%	77%	96.6	14%	55%	287.3
Vietnam	49%	89%	74.4	22%	61%	166.4
Cambodia	12%	64%	421.8	0%	29%	–
Laos	0.2%	47%	21,918.2	0%	47%	–
Ethnicity (Born in French Indochina)						
Vietnamese	56%	92%	58.1	19%	61%	207.5
Cambodian	14%	62%	335.7	0%	22%	–
Chinese	34%	86%	141.3	36%	92%	150.8
Laotian	2%	54%	2628.9	5%	54%	872.4
Hmong	0%	0%	–	0%	28%	–
Developing-country Asian immigrants						
China, India, Korea, Philippines	53%	108%	96.4	46%	94%	102.4

Source: Authors' estimates. For the 1975–1980 cohort, entry earnings estimates are based on a 6% microdata sample combining the 1980 Census of Population Public use 5% and 1% samples; earnings ten years later are based on a 6% sample combining the 1990 Census of Population 5% and 1% samples. For the 1985–1990 cohort, entry earnings estimates are based on a 6% microdata sample combining the 1990 Census of Population 5% and 1% samples; earnings ten years later, for the 1985–1990 cohort, are based on 2000 Census of Population 5% and 1% samples. Appendix A provides information on sample sizes for entry cohorts at entry and ten years later

Notes: Foreign born are defined as persons born outside of the United States excluding those with U.S. parents. Groups are defined by country-of-origin information or ethnicity as defined by responses to the first ancestry code. The sample does not exclude students, the self-employed, and persons with zero earnings

Table 21.6 Mean entry earnings, earnings ten years later, and ten-year growth rates for the 1975–1980 and 1985–1990 entry cohorts; Earnings at entry and ten years later are expressed as a percentage of U.S. native-born mean earnings

	1975–1980 cohort			1985–1990 cohort		
	Ages 25–54	Ages 35–64		Ages 25–54	Ages 35–64	
	1979 earnings	1989 earnings	10-year growth rate	1989 earnings	1999 earnings	10-year growth rate
Place of birth						
Indochina	42%	76%	92.7	27%	53%	115.3
Vietnam	47%	85%	91.4	30%	59%	114.3
Cambodia	27%	61%	145.8	22%	43%	115.2
Laos	23%	40%	83.7	22%	41%	106.5
Ethnicity (born in French Indochina)						
Vietnamese	51%	86%	80.6	30%	59%	114.3
Cambodian	29%	61%	124.9	20%	39%	112.8
Chinese	45%	93%	120.5	42%	96%	151.0
Laotian	23%	45%	105.4	25%	48%	106.0
Hmong	20%	24%	30.7	8%	30%	333.1
Developing-country Asian immigrants and US born						
China, India, Korea, Philippines	60%	117%	108.1	52%	102%	116.1
U.S. Native-born	29,308	31,140	6.3	29,293	32,161	9.8

Source: Authors' estimates. For the 1975–1980 cohort, entry earnings estimates are based on a 6% microdata sample combining the 1980 Census of Population Public use 5% and 1% samples; earnings ten years later are based on a 6% sample combining the 1990 Census of Population 5% and 1% samples. For the 1985–1990 cohort, entry earnings estimates are based on a 6% microdata sample combining the 1990 Census of Population 5% and 1% samples; earnings ten years later, for the 1985–1990 cohort, are based on 2000 Census of Population 5% and 1% samples. Appendix A provides information on sample sizes for entry cohorts at entry and ten years later

Notes: Foreign born are defined as persons born outside of the United States excluding those with U.S. parents. Groups are defined by country-of-origin information or ethnicity as defined by responses to the first ancestry code. The sample does not exclude students, the self-employed, and persons with zero earnings

entering immigrants plummeted. There is, however, no inter-cohort change in their school attendance rate. The educational achievement of entering Laotians and Cambodians stayed fairly constant across the cohorts, and school attendance increased. By ethnicity, there are increases in school attendance for all groups other than the Vietnamese. These comparisons tentatively suggest that, holding source country schooling levels constant, the propensity to attend school for the 1985–1990 versus 1975–1980 cohort of Indochinese men increased.

Concluding Remark

Indochinese refugees of the 1985–1990 cohort arrived in the United States in a severely disadvantaged state: their disability rates were quadruple those of other Asian immigrants, their public assistance rate (for working-age men) 32 times those of Asian immigrants from economically

Table 21.7 Comparison of the school attendance during the first five years of the 1975–1980 and 1985–1990 entry cohorts: Immigrant men, 25–64 years old

	1975–1980 cohort	1985–1990 cohort
	1980	1990
Place of birth in South East Asia		
Vietnam	26%	26%
Cambodia	21%	25%
Laos	18%	33%
Ethnicity (born in Indochina)		
Vietnamese	26%	26%
Cambodian	19%	26%
Chinese	20%	25%
Laotian	16%	27%
Hmong	24%	49%
Non-South East Asian comparison		
China, India, Korea, Philippines	19%	28%
Western Europe	10%	18%

Source: Authors' estimates based on a 6% 1980 file created by combining and reweighting the 5% and 1% files of the 1980 Census of Population and a 6% microdata 1990 sample created by combining and reweighting the 1990 Census of Population Public Use 5% and 1% samples. Appendix A provides information on sample sizes by entry cohort

developing countries, and their schooling levels woefully low. Yet, Indochinese refugees—including those who endured years in refugee camps—resemble the developing country Asian immigrants in one key aspect: their extraordinary school attendance rates.

Reference

Reimers David M. 2005. *Other Immigrants: The Global Origins of the American People*, NY: NYU Press.

PART VI

A Brief Glance Backward and Conclusion

In thinking about the immigration mosaic that is America, Asians are often viewed as new arrivals even though their U.S. settlement began well before 1900. This chapter goes back in time to America's first Asian immigrants and examines the educational and economic attainment of their descendants.

CHAPTER 22

A Longer Perspective on Initial Conditions and Immigrant Adjustment

Documents from the former Immigration and Naturalization Service (INS) give a historical record of the number of immigrants by country and the occupations they held in their countries of origin. The number of immigrants reporting no occupation is also recorded. We compiled these records from 1870 through 1980 so that the occupational backgrounds of immigrants from each country could be traced over time. With information on the average skill levels of entering immigrants, they allow comparisons of the human capital background of Asian immigrants from the earliest entrants to the most recent. They also provide a baseline with which to gauge achievements of the descendants of the first wave of Asian immigrants.

The educational backgrounds of Indochinese refugees, described in Part V, contrast sharply with the schooling levels of the post-1965 Asian immigrants. The occupations immigrants held in their original countries, as recorded in the Immigration and Naturalization Service (INS) annual reports, further highlight this contrast. Among immigrants entering the United States in 1976–1980, who reported an occupational background, nearly half from China, Japan, and the Philippines, and three-quarters from India and Korea, report a professional occupation prior to migrating versus 15.5% of Vietnamese entrants.[1]

---

[1] We compiled these data from the annual reports of the former INS.

Table 22.1 Percentage laborers of working immigrants (The numbers in parentheses are the total admitted who report an occupation)

	Chinese	Japanese	Korean	Indian	Filipino	
(A) The early years: 1871–1935						
1871–1875	93.0% (16,437)					
1876–1880	95.1% (57,773)					
1881–1885	96.9% (59,779)					1882: Chinese Exclusion Act limited Chinese laborers
1891–1895	0.0% (13,384)					
1896–1900	17.9% (8322)	45.4% (19,779)				
1901–1905	30.1% (12,537)	51.0% (65,148)	95.7% (7475)			
1906–1910	4.0% (7129)	75.9% (67,558)		85.4% (5172)		1907: Gentlemen's agreement with Japan-limited Japanese laborers
1911–1915	7.8% (9760)	60.2% (36,599)				
1916–1920	5.4% (9505)	51.6% (47,139)				
1921–1925	12.2% (18,947)	35.5% (28,707)		88.7% (11,944)		1924: Immigration Act—Origins of national quota system
1926–1930	24.5% (5398)	2.6% (3292)		88.7% (42,803)		
1931–1935				88.6% (5947)		1934: Philippines granted commonwealth status
(B) The second wave: 1966–1980						
1966–1970	2.5% (75,748)	4.9% (19,395)	1.0% (25,618)	0.7% (27,859)	8.3% (85,636)	1965: Immigration Act dropped quotas; anti-Asian bias ended

(*continued*)

Table 22.1 (continued)

	Chinese	Japanese	Korean	Indian	Filipino
1971–1975	4.4%	3.4%	1.7%	0.9%	4.9%
	(85,645)	(23,809)	(112,493)	(72,912)	(153,254)
1976–1980	9.4%	2.9%	4.3%	4.2%	9.6%
	(84,166)	(16,494)	(120,256)	(76,561)	(154,908)

Source: Authors' estimates based mostly on Immigration and Naturalization Service annual reports. See Appendix B for information on their construction

Using the INS records, we can compare the occupational backgrounds of Asian immigrants who entered the United States between 1870 and 1924 (Table 22.1 top half), with those of the post-1965 Asian immigrants (Table 22.1 bottom half).[2] Purely in terms of pre-migration human capital levels, the Indochinese refugees studied in Part V have more in common with America's earliest Asian entrants than with the post-1965 Asian immigrants, who were the focus of Parts II through IV.[3]

The First Wave of Asian Immigration: Entry Characteristics

Before 1890, Asian immigration hailed almost exclusively from China. This immigration started around 1850 and peaked in the 1870s, when more than 123,000 Chinese immigrants were recorded. During the peak year of Chinese immigration, close to 97% of those with a recorded occupational background were unskilled laborers in China (Table 22.1).

With the Chinese Exclusion Act, Chinese immigration dropped from 39,579 in 1882 to 279 two years later.[4] Concomitantly, Indians, Koreans, and particularly Japanese, filled the demand for inexpensive unskilled labor. Like their Chinese predecessors, these immigrants generally worked as unskilled laborers in their countries of origin. Although the occupational backgrounds of Japanese immigrants were more diverse than those

---

[2] Appendix B describes how we used the INS annual reports to create the series in Table 21.1.

[3] Please refer to "Asians in Hawaii and the United States," pp. 40–70 or Reimers (2005).

[4] Although the immigration of Chinese laborers was strictly limited, other groups of Chinese, such as merchants and later, the wives and families of Chinese Americans, continued to immigrate.

of the Chinese, fully 75% of working Japanese entrants reported laborer as their occupational background during the peak year of their U.S. immigration. During the peak years of Korean and Indian immigration, 96% of Korean immigrants and 85% of Indian immigrants (with recorded occupations) reported laborer as their country-of-origin vocation (Table 22.1).

Following the 1907 Gentlemen's Agreement, Japanese immigration dropped from 30,824 in 1907 to 3275 in 1909, rebounding later during the years 1911–1920 to about two-thirds its level before the agreement.[5] The Immigration Act of 1924 (the Johnson-Reid Act) declared that all aliens ineligible for citizenship were also ineligible for immigration. Since foreign-born Asians did not qualify for citizenship, being neither "free whites" or of "African descent,"[6] this law effectively barred Asians from immigrating to the United States.[7] The 1924 law did not, however, restrict immigration from the Philippine Islands which, at this time, were U.S. territory.[8] Demand for Filipino labor rose. Between 1921 and 1932, an average of 5000 Filipinos entered each year, reaching a peak of 11,360 in 1929. Like the other early Asian immigrant groups, the most common occupational background of entering Filipino immigrants was unskilled laborer: 89% reported laborer as their occupational background (Table 22.1). When the Philippines attained commonwealth status in 1934, Filipinos became aliens for immigration purposes; their annual immigration quota was set at 56 persons. With the end of Filipino immigration, Asian immigration all but ceased, a state that persisted for 30 years.

---

[5] The 1907 Gentlemen's Agreement between the U.S. and Japan gave the Japanese the right to select immigrants and, as a result, severely limited Korean immigration, as Korea was controlled by Japan at that time.

[6] The Naturalization Act of 1790 granted the right to apply for citizenship to "free white persons." Citizenship privileges were extended to "aliens of African nativity and persons of African descent" in 1870.

[7] Immigration for Asians other than Filipinos was limited to wives and dependent children of American-born Asians or Asians who had become citizens before the anti-Asian restrictions were in place.

[8] The Treaty of Paris of 1899 transferred possession of the Philippine Islands from Spain to the U.S.

Their Descendants

The virtual ban on Asian immigration from 1924 to 1965 meant that in 1960, U.S.-born Asians were mostly the descendants of immigrants who had entered the United States between 1870 and 1924. Analyzing the socioeconomic status of U.S.-born Asians in 1960 provides a picture of intergenerational progress almost unadulterated by new immigrant flows.

Comparing the schooling levels of U.S.-born men reveals that, in 1960, Asian education levels generally surpassed those of non-Hispanic whites.[9] While U.S.-born non-Hispanic white men (ages 25–64) averaged 10.7 years of schooling, U.S.-born Chinese and Japanese men averaged 11.9 and 11.7 years, respectively; 27.3% of Chinese Americans and 15.1% of Japanese Americans were college graduates in 1960, compared with 11.7% of U.S.-born non-Hispanic white men. U.S.-born Filipino men were less educated than the other Asian American groups. Yet, with an average of 10.23 years of schooling and with 6.3% college graduates, they were not far behind the 1960 non-Hispanic white benchmark.

In earnings and wages, U.S.-born Chinese and Japanese men also compared favorably in 1960 with U.S.-born non-Hispanic white men. On average, Japanese men earned 8% more than the U.S.-born non-Hispanic whites and Chinese men earned 12% more. Measured in hourly wages, Japanese men earned 1% more and Chinese men 7% more than U.S.-born non-Hispanic whites. Filipino men earned 73% of what U.S.-born white men earned in 1960 and had wages that were 83% as high.

These statistics attest to the fact that by 1960 the sons and grandsons of the early twentieth-century Asian immigrants had made considerable economic progress from the laborer roots of their forefathers. While early Asian migrants were low skilled, the typical U.S.-born Asian in 1960 earned on par with the typical U.S.-born non-Hispanic white. Their success by 1960 was largely attributable to high levels of educational investment, an investment even more remarkable given the discrimination U.S.-born Asians faced in 1960 in the labor market (Duleep and Sanders 2012).

The Asian story of intergenerational educational and economic growth related in this chapter's brief analysis echoes studies of European intergenerational achievement in the United States.[10] It extends to an

---

[9] When we first did this analysis, the U.S. Census Bureau had released only a 1% sample of the 1960 long form data. It is thus restricted to the three most populated Asian groups in 1960: Chinese, Japanese, and Filipinos.

[10] Relevant studies include Duleep and Sider (1986, and studies referenced therein), Lieberson (1996a, b and studies referenced therein, 1980) and Thernstrom (1973).

intergenerational arena an important theme of this book—immigrant entry earnings are a flawed predictor of immigrant economic success—and resonates with an observation by Lieberson (1996b, p. 343) concerning European immigration to the United States.

> Immigration history in the United States provides an important lesson. Namely, there is little justification for using the initial problems faced by the immigrants to draw pessimistic long-term conclusions about either the group or American society.

REFERENCES

Duleep, H. and Sanders, S. (2012) The Economic Status of Asian Americans Before and After the Civil Rights Act, IZA Discussion Paper No. 6639

Duleep, H. and Sider, H. (1986) *The Economic Status of Americans of Southern and Eastern European Ancestry*, U.S. Commission on Civil Rights, http://www.law.umaryland.edu/marshall/usccr/documents/cr11089z.pdf

Lieberson S. 1980. *A Piece of the Pie: Blacks and White Immigrants Since 1880.* Berkeley, Calif.: University of California Press 1980;

Lieberson S. 1996a. "Earlier Immigration to the United States: Historical Clues for Current Issues of Integration." In: Carmon N Immigration and Integration in Post-Industrial Societies. London; New York: Macmillan Press Ltd; St Martin's Press; pp. 187–205.

Lieberson S. 1996b. "Contemporary Immigration Policy: Lessons from the Past." In: Duleep HO, Wunnava PV Immigrants and Immigration Policy: Individual Skills, Family Ties, and Group Identities. Greenwich, CT: JAI Press; pp. 335–351.

Reimers David M. 2005. *Other Immigrants: The Global Origins of the American People*, NY: NYU Press.

Thernstrom, Stephan, (1973) (Reprint 2014) The Other Bostonians: Poverty and Progress in the American Mstropolis, 1880–1970, Harvard University Press, Cambridge, MA and London, England.

CHAPTER 23

Conclusion

The demise of a national-origin quota system for U.S. immigration, and its replacement in 1965 with an admission policy emphasizing family reunification, opened the gates to an increasing flow of immigrants differing dramatically in country-of-origin composition from earlier U.S. immigration. Asian immigration was an important part of this development (Fig. 1.1).

Various concerns have been voiced about the "new," primarily Asian and Hispanic, immigrants. They hail from homelands less economically developed than the United States, often with less equal income distributions, and perceived to be more culturally, institutionally, and linguistically distinct from the United States than the West European homelands of yesteryear's immigrants. The new immigrants' entry earnings are also low relative to earlier cohorts from the same countries, and relative to immigrants from economically developed countries. Moreover, the decline in entry earnings persists adjusting for educational attainment. This finding has prompted scholars to conclude that high kinship-based admissions and the shift in country of origin for U.S. immigrants created a decline in immigrant labor market quality.

Yet, if entry earnings are correlated with any factor that enables or increases human capital investment then immigrant entry earnings are not a good measure of either human capital or unmeasured immigrant quality: home-country skills and attributes that do not yield an immediate labor market return may still represent human capital in a meaningful way if they complement or aid in the acquisition of U.S. skills. Family-based immigrants and refugees do not enjoy the immediate high demand for their skills that, by definition, employment-based immigrants do. But they experience much higher earnings growth. Moreover, for all of the non-refugee groups from economically developing countries analyzed in this book, the low adjusted earnings of immigrant men during their initial U.S. years are offset by the earnings of immigrant women.

THE IMMIGRANT HUMAN CAPITAL INVESTMENT MODEL

The analyses of this book suggest that the post-1965 decline in immigrant entry earnings occurred because of a decline in immigrant skill transferability. The crux of the Immigrant Human Capital Investment (IHCI) Model (Chap. 4) is that as the transferability of source-country human capital falls, its value in the labor market falls more than its value in learning new human capital. Thus, holding level of source-country human capital constant, immigrants with low skill transferability will invest more in host-specific human capital than immigrants with high skill transferability. Most importantly, this higher incentive to invest in human capital extends beyond U.S.-specific human capital, such as English proficiency, that restores the value of source-country human capital to new human capital in general.

A methodological ramification of the IHCI Model is that when earnings differences stem from differences in the ease with which foreign skills transfer to the United States, earnings growth will (adjusting for initial levels of human capital) be inversely related to entry earnings. Ignoring the inverse relationship between entry earnings and earnings growth understates immigrant earnings growth when the education-adjusted entry earnings of immigrants fall over time and overstates earnings growth when entry earnings increase. Ideally, analyses of immigrant earnings trajectories should let entry earnings and earnings growth freely vary or, in parametric models, include variables (such as immigrant admission criteria) to capture cohort effects and allow those variables to interact with entry earnings *and* earnings growth (Chap. 8). Sample exclusions that labor economists commonly impose, as well as other practices, further work to understate earnings growth for immigrants with low initial earnings in analyses that follow cohorts as opposed to individuals (Chap. 5).

In keeping with the predictions of the IHCI model, men from Asian developing countries start their U.S. lives with earnings substantially below the earnings of West European and Canadian immigrants or U.S. natives. Yet, their earnings growth is much higher (Chaps. 6, 7, and 17).

The groups with high earnings growth invest heavily in human capital as evidenced by high school attendance rates, even at ages 35 and over (Chaps. 9, 17). Such investment occurs even among very poorly educated Indochinese refugees who suffered the scars of war and extended periods in refugee camps (Chaps. 19, and 21). Indeed, school attendance among Indochinese refugees exceeds the school attendance of non-refugee Asian immigrants, which exceeds the school attendance of West European immigrants.

Among other things, these findings suggest another potential source of bias in almost all work by economists on immigrant earnings. When measuring immigrant earnings growth in a regression, economists generally control for level of education using post-immigration educational achievement as opposed to the education level immigrants had when they entered the United States. This understates the earnings growth of all immigrants, relative to the U.S. born, but particularly that of (permanent) immigrants initially lacking U.S.-specific human capital (Chap. 5).

Our statistical analyses, focused on individual attributes, reveal trends and correlations that broadly support the IHIC model but also highlight distinctive group-specific patterns that scholars in fields other than economics are better equipped to elucidate.[1]

THE IMPORTANCE OF BEING PERMANENT

A theme throughout our book is the importance of being permanent. We propose that the decision to invest in U.S.-specific human capital is jointly determined with the decision to stay in the United States. Immigrants would embark on investments such as starting a business, taking jobs with on-the-job training, and learning English only if the benefits from investing could be reaped in the future. Immigrants who do not plan on staying in the United States are likely to be persons who can work here without

---

[1] For examples of this more in depth approach see, for instance, Glenn (1983), Hirschman and Wong (1986), Lieberson and Waters (1988), Min (2008), Nee et al. (1994), Portes (1995), Portes and Rumbaut (1996), Reimers (2005), Rumbaut (1994), Sanders and Nee (1987), Thernstrom (1973), Waters and Eschbach (1995), Waters and Lieberson (1992), and White et al. (1993).

investing, either because their U.S. jobs do not require U.S.-specific skills (as we hypothesized for a majority of Japanese immigrant men), or because their country-of-origin skills are generally highly transferable to the U.S. labor market.

We find that groups with relatively low adjusted entry earnings and high earnings growth are more likely to be permanent, as measured by emigration and naturalization rates, than groups with relatively high adjusted entry earnings (Chap. 10). From a perspective of boosting human capital investment, policies to promote immigrant permanence should be encouraged. Temporary migration, whose temporary nature is voluntary, has an increasingly important role to play in a global economy, but policies that purposely discourage long-term migration are short-sighted.

Policy Implications of Our Results

Policy makers have viewed the U.S. post-1965 emphasis on family admissions solely from a humanitarian perspective. Our research suggests economic benefits: immigrants who come into the United States via family admissions have higher earnings growth, hence higher human capital investment, than employment-based immigrants (Chap. 8). Sibling admissions are positively associated with earnings growth and, among Asians, a high propensity for self-employment (Chap. 16).

An important implication of the inverse relationship between entry earnings and earnings growth is that entry earnings poorly predict economic success and should not be part of the policy debate. Human capital investment by immigrants should be central to all projections of the economic contributions and burdens resulting from immigration. Conditional on permanence, the IHCI Model predicts a high rate of human capital investment and earnings' growth for immigrants with low initial skill transferability. Low initial skill transferability would characterize most immigrants from the lesser developed countries of Asia and Africa as well as most Latin American immigrants—together a large majority of new migrants to the United States.

The optimality of high human capital investment for most immigrants facilitates educational investment and occupational change. This has the important policy implication that recent immigrants may be better equipped than U.S. natives or immigrants with high skill transferability to dynamically respond to the changing skill needs of the U.S. economy.

Immigrants are more occupationally mobile even long after arrival, indicating immigration may contribute to a more flexible labor force (Green 1999)

The well-publicized concern about a decline in immigrant labor market quality grew from methodological flaws in the analysis of immigrant earnings (Chap. 5). Our research shows that lower initial earnings have been accompanied by higher earnings growth. Higher earnings growth is inconsistent with lower labor market quality.

Recognizing the importance of post-migration human capital investment puts the post-1965 immigration in a positive light. The immigrants who entered the United States after 1965 are not of lower "quality" than earlier immigrants, but are meeting the challenges of a U.S. labor market that is more selective about skills through melding high rates of post-migration human capital investment to their original skills. Rather than contributing to an underclass, they have become the most upwardly mobile of American workers. This greatly benefits both immigrants and the U.S. economy.

References

Glenn, EN. 1983. "Split Household, Small Producer and Dual Wage Earner: An Analysis of Chinese-American Family Strategies," *Journal of Marriage and the Family*, 45(1): 35–46.

Green, D. (1999). "Immigrant Occupational Attainment: Assimilation and Mobility over Time." *Journal of Labor Economics*, 17(1): 49–79.

Hirschman C, Wong MG. 1986. "The Extraordinary Educational Attainment of Asian-Americans: A Search for Historical Evidence and Explanations." *Social Forces* 65 (1): 1–27.

Lieberson S, Waters MC. 1988. *From Many Strands: Ethnic and Racial Groups in Contemporary America*. New York: Russell Sage Foundation.

Min, Pyong Gap. 2008. *Ethnic Solidarity for Economic Survival: Korean Greengrocers in New York City*, NY: Russell Sage Foundation.

Nee V, Sanders J, Sernau, S. 1994. "Job Transitions in an Immigrant Metropolis: Ethnic Boundaries and the Mixed Economy," *American Sociological Review*, 59(6): 849–872.

Portes A. 1995. "Economic Sociology and the Sociology of Immigration: A Conceptual Overview," in *The Economic Sociology of Immigration: Essays on Networks, Ethnicity, and Entrepreneurship*, A. Portes, ed. New York: Russell Sage Foundation.

Portes A and Rumbaut RG. (1996) *Immigrant America: A Portrait* (2nd ed.). Berkeley, Calif.: University of California Press.

Reimers David M. 2005. *Other Immigrants: The Global Origins of the American People*, NY: NYU Press.

Rumbaut, R.G. (1994) "Origins and Destinies: Immigration to the United States since World War II," *Sociological Forum* 9(4):583–621.

Sanders JM, Nee V. 1987. "Limits of Ethnic Solidarity in the Enclave Economy." *American Sociological Review* 52 (6): 745–773.

Thernstrom, Stephan, (1973) (Reprint 2014) *The Other Bostonians: Poverty and Progress in the American Mstropolis, 1880–1970,* Harvard University Press, Cambridge, MA and London, England.

Waters, MC and Karl Eschbach (1995) "Immigration and Ethnic and Racial Inequality in the United States," *Annual Review of Sociology*, 21: 419–446.

Waters MC, Lieberson S. 1992. "Ethnic Differences in Education: Current Patterns and Historical Roots," in *International Perspectives on Education and Society* (Vol. 2), A Yogev, ed. Greenwich, Conn.: JAI Press; pp. 171–187.

White MJ, Biddlecom AE, Gou S. 1993. "Immigration, Naturalization, and Residential Assimilation among Asian Americans in 1980." *Social Forces* 72 (91): 93–117.

Appendix A: Sample Size Information for Year-of-Entry Cohorts at Entry and Ten Years Later by Age and Education Categories

Table A.1 Unweighted sample sizes for 1965–1970 entry cohort on 1970 (six 1% files) and 1980 (5% and 1% files) census PUMS: Males, high school or less

	Age in 1970		Age in 1980			
	25–39	40–54	25–54	35–49	50–64	35–64
China	179	162	341	271	187	458
India	62	14	76	43	20	63
Japan	76	19	95	39	11	50
Korea	20	6	26	19	6	25
Philippines	233	99	332	219	106	325
Europe	2832	1666	4498	2794	1473	4267
Western Europe	2382	1314	3696	2257	1139	3396
English-speaking Western Europe	406	200	606	245	134	379
Non-English-speaking Western Europe	1976	1114	3090	2012	1005	3017
Eastern Europe	450	352	802	537	334	871

Source: Authors' estimates

Note: On the three 1970 1% Census Files with no year of immigration data, residing abroad five years before is used as an indicator of being in the 1965–1970 cohort

© The Author(s), under exclusive license to Springer Nature Switzerland AG 2020
H. Duleep et al., *Human Capital Investment*,
https://doi.org/10.1007/978-3-030-47083-8

Table A.2 Unweighted sample sizes for 1965–1970 entry cohort on 1970 (six 1% files) and 1980 (5% and 1% files) census PUMS: Males, more than high school

	Age in 1970			Age in 1980		
	25–39	40–54	25–54	35–49	50–54	35–64
China	464	75	539	565	77	642
India	702	79	781	810	72	882
Japan	358	63	421	91	11	102
Korea	251	22	273	270	29	299
Philippines	665	140	805	774	161	935
Europe	1958	549	2507	1632	490	2122
Western Europe	1629	370	1999	1237	315	1552
English-speaking Western Europe	690	179	869	530	143	673
Non-English Western Europe	939	191	1130	707	172	879
Eastern Europe	329	179	508	395	175	570

Source: Authors' estimates

Note: On the three 1970 1% Census Files with no year of immigration data, residing abroad five years before is used as an indicator of being in the 1965–1970 cohort

Table A.3 Unweighted sample sizes for 1975–1980 entry cohort on 1980 census (5% and 1% files) and 1990 census (5% and 1% files): Males, high school or less

	Age in 1980			Age in 1990		
	25–39	40–54	25–54	35–49	50–64	35–64
China	477	354	831	475	297	772
India	286	152	438	217	90	307
Japan	162	37	199	70	10	80
Korea	377	213	590	248	161	409
Philippines	278	145	423	276	127	403
Europe	1526	847	2373	1411	670	2081
Western Europe	1292	655	1947	1026	438	1464
English-speaking Western Europe	223	87	310	161	69	230
Non-English Western Europe	1069	568	1637	865	369	1234
Eastern Europe	234	192	426	385	232	617

Source: Authors' estimates

Table A.4 Unweighted sample sizes for 1975–1980 entry cohort on 1980 census (5% and 1% files) and 1990 census (5% and 1% files): Males, more than high school

	Age in 1980		Age in 1990			
	25–39	40–54	25–54	35–49	50–54	35–64
China	1173	277	1450	1000	252	1252
India	1609	268	1877	1343	242	1585
Japan	802	199	1001	177	26	203
Korea	808	297	1105	684	272	956
Philippines	1151	277	1428	1190	281	1471
Europe	2665	989	3654	1875	657	2532
Western Europe	1873	527	2400	1262	306	1568
English-speaking Western Europe	826	253	1079	647	162	809
Non-English Western Europe	1047	274	1321	615	144	759
Eastern Europe	792	462	1254	613	351	964

Source: Authors' estimates

Table A.5 Unweighted sample sizes for 1985–1990 entry cohort on 1990 (5% and 1% files) and 2000 census (5% and 1% files): Males, high school or less

	Age in 1990		Age in 2000			
	25–39	40–54	25–54	35–49	50–64	35–64
China	798	506	1304	1157	607	1764
India	589	222	811	891	291	1182
Japan	166	69	235	73	27	100
Korea	430	291	721	438	254	692
Philippines	402	182	584	475	300	775
Europe	2011	873	2884	2154	860	3014
Western Europe	1119	370	1489	1090	373	1463
English-speaking Western Europe	433	111	544	338	94	432
Non-English Western Europe	686	259	945	752	279	1031
Eastern Europe	892	503	1395	1064	487	1551

Source: Authors' estimates

Table A.6 Unweighted sample sizes for 1985–1990 entry cohort on 1990 (5% and 1% files) and 2000 census (5% and 1% files): Males, more than high school

	Age in 1990			Age in 2000		
	25–39	40–54	25–54	35–49	50–54	35–64
China	2447	669	3116	2590	661	3251
India	2199	449	2648	2894	551	3445
Japan	1244	471	1715	255	84	339
Korea	1187	383	1570	967	315	1282
Philippines	1426	584	2010	2023	735	2758
Europe	3696	1415	5111	3897	1331	5228
Western Europe	2528	813	3341	2037	544	2581
English-speaking Western Europe	1111	427	1538	990	288	1278
Non-English Western Europe	1417	386	1803	1047	256	1303
Eastern Europe	1168	602	1770	1860	787	2647

Source: Authors' estimates

Table A.7 Unweighted sample sizes for 1995–2000 entry cohort on 2000 census (5% and 1% files) and 2010 American community survey: Males

	Age 25–54 in 2000	Age 35–64 in 2010
China	5046	1039
Mainland	3810	866
Hong Kong	361	56
Taiwan	875	117
India	8473	1685
Japan	1997	79
Korea	2585	352
Philippines	2299	535

Source: Authors' estimates

Note: in the analyses of the book, we increased the sample size by using the 1% files for 1998, 1999, and 2000. 10% samples are now available as of this writing

APPENDIX A: SAMPLE SIZE INFORMATION FOR YEAR-OF-ENTRY... 261

Table A.8 Unweighted sample sizes for Southeast Asian refugee men: The 1975–1980 entry cohort, males, high school or less: 1980 census (5% and 1% files) and 1990 census (5% and 1% files)

	Age in 1980		*Age in 1990*			
	25–39	40–54	25–54	35–49	50–64	35–64
Indochina	1249	509	1758	896	424	1320
Cambodia	127	38	165	102	37	139
Laos	285	96	381	208	82	290
Vietnam	837	375	1212	586	305	891

Source: Authors' estimates

Table A.9 Unweighted sample sizes for Southeast Asian refugee men: The 1975–1980 entry cohort, males, more than high school: 1980 census (5% and 1% files) and 1990 census (5% and 1% files)

	Age in 1980		*Age in 1990*			
	25–39	40–54	25–54	35–49	50–54	35–64
Indochina	1146	378	1524	1146	343	1489
Cambodia	72	14	86	85	17	102
Laos	97	25	122	106	19	125
Vietnam	977	339	1316	955	307	1262

Source: Authors' estimates

Table A.10 Unweighted sample sizes for Southeast Asian refugee men: 1985–1990 entry cohort on 1990 (5% and 1% files) and 2000 census (5% and 1% files): Males, high school or less

	Age in 1990		*Age in 2000*			
	25–39	40–54	25–54	35–49	50–64	35–64
Indochina	1111	531	1642	1377	716	2093
Cambodia	158	90	248	211	94	305
Laos	271	109	380	351	158	509
Vietnam	682	332	1014	815	464	1279

Source: Authors' estimates

Table A.11 Unweighted sample sizes for Southeast Asian refugee men: 1985–1990 entry cohort on 1990 (5% and 1% files) and 2000 census (5% and 1% files): Males, more than high school

	Age in 1990		Age in 2000			
	25–39	40–54	25–54	35–49	50–54	35–64
Indochina	480	220	700	708	262	970
Cambodia	56	26	82	73	14	87
Laos	92	44	136	89	18	107
Vietnam	332	150	482	546	230	776

Source: Authors' estimates

Table A.12 Foreign-born women married to foreign-born men, the 1975–80 cohort: Unweighted sample sizes for 1975–1980 entry cohort on 1980 census (5% and 1% files) and 1990 census (5% and 1% files)

	Age in 1980	Age in 1990
	25–54	35–64
China	1636	1568
India	1627	1398
Japan	789	153
Korea	1412	1234
Philippines	1281	1634
Indochina	1915	1855
Cambodia	143	121
Laos	336	270
Vietnam	1436	1464
Europe	3997	2623
Western Europe	2420	1470
English-speaking Western Europe	754	408
Non-English Western Europe	1666	1062
Eastern Europe	1577	1153

Source: Authors' estimates

APPENDIX B: NOTES ON HISTORICAL DATA IN CHAP. 22

The INS occupational background records do not distinguish between men and women, or between persons of working age and children and the retired. In using these statistics, it is important to distinguish the labor force from persons outside the labor force. Otherwise, a decrease in the percentage of immigrants reporting laborer occupations, for instance, could simply reflect an increase in the number of women and children in the immigrant group, rather than an increase in the occupational backgrounds of the immigrant work force. We estimated the number of persons in the labor force for each year as: LF = (total number of immigrants) − (number reporting no occupation). We then estimated the percentage of the entering immigrant labor force reporting a particular occupation for each year as: (number of persons reporting occupation i)/LF.

These numbers, averaged over five-year periods, are presented in Table 22.1. The category "laborer" includes the INS category laborer as well as the category farm laborer. The category "professional" includes the INS composite category professional as well as the category merchant and manager.

Statistics on Filipino Immigrants Before 1935
Before 1935 Filipinos were not considered immigrants. We derived statistics on the total number of Filipino entrants for the years 1911–1932 from the following INS reports and locations: 1911–1920—Table 110 of the *Report of the Commissioner General of Immigration*, p. 260;

© The Author(s), under exclusive license to Springer Nature Switzerland AG 2020
H. Duleep et al., *Human Capital Investment*,
https://doi.org/10.1007/978-3-030-47083-8

1921–1930—Table 111, note 1, of the 1929 and 1930 *Reports* and from Tables 110 and 111 of the 1931 *Report*; 1931–1940—Tables 110 and 111 of the 1931 *Report* and Table 64, note 1, of the 1932 *Report*.

We derived information on the percentage of Filipinos who were laborers from data presented in Honorante Mariano, *The Filipino Immigrants in the United States*, doctoral dissertation, University of Oregon, 1933 (reprinted in 1972 by R and E Research Associates, publishers and distributors of ethnic studies). To estimate the percentage of the Filipino immigrant labor force that was laborers, we assumed that females and males under 16 years of age had no occupation. Several figures suggest that the male-to-female ratio was 15:1 (see Mariano dissertation, above, and U.S. census, 1930, vol. III, pts. I and II) and that about 3% of males were under 16 years of age (Mariano dissertation, above, pp. 21, 22, derived from State of California records on Filipino entrants).

The number of Filipino immigrants with no occupation was estimated for each year as:

$NOOCC = 1/15$ (total immigrants) $+ 3/100$ ($4/15$[total immigrants])
and $LF = $ total immigrants $- NOOCC$.

Index[1]

A

Admission
 family-based, 81, 83, 92
 refugee, 5, 5n4
Anti-Chinese sentiment, 20, 21
Asia, vii, 1, 9, 9n6, 19–24, 31, 31n5, 58, 85n9, 86, 88, 108, 254
Asian countries, 3, 9, 81n2
 developing, 73, 73n14, 76
Asian groups, 9, 11, 12n11, 65–77, 95, 96n3, 122–132, 122n3, 137, 138, 140–143, 145–147, 152, 170, 171, 197, 199, 249n9
Asian immigrants, 1, 2n1, 3, 9, 11, 12, 24, 42, 53, 56, 58, 61, 62, 65, 67, 72, 74n14, 82, 83, 85, 92, 95n1, 98, 100, 103, 108, 112, 117, 122, 123, 127n14, 138n7, 143, 159, 160, 162, 163, 180, 180n5, 183, 184, 195–197, 204, 211, 213, 214, 216, 217n1, 218–221, 229, 234, 236, 238, 241–243, 245, 247–249, 253
Asian immigration, 1, 4, 5, 9, 11, 19, 21, 22, 31n5, 65, 67n5, 82, 83, 199, 247–249, 251
Asian Indians, 24, 73, 138, 138n7, 180n6

B

British immigrants, 72, 81, 83, 88, 107

C

California, 19–22, 24, 202, 264
Cambodian, 199, 204, 206, 212, 213, 218, 223, 223n1, 225, 227–229, 231, 232, 235–237, 241
Canadian Families, 125, 128, 128n15

---

[1] Note: Page numbers followed by 'n' refer to notes.

© The Author(s), under exclusive license to Springer Nature Switzerland AG 2020
H. Duleep et al., *Human Capital Investment*,
https://doi.org/10.1007/978-3-030-47083-8

265

Canadian Married Immigrant
  Women, 147
Canadians, 65–67, 67n5, 70, 71, 74,
  75, 107, 108, 117, 121n2,
  122–133, 127n14, 137, 138,
  140–143, 145–148, 152, 153,
  157, 159, 162, 163, 170, 171,
  174, 195, 200
  immigrant families, 127n14
  immigrants, 54, 108, 112, 132,
    143, 170, 195, 253
  immigrant women, 138, 141, 142n9
  women, 131, 141
Census race variable, 122, 122n3,
  124–132, 137, 140–142,
  145–147, 152, 170, 171
Census samples, 49, 50, 56, 84n4,
  85n9, 86, 87n10, 88, 90, 91,
  109, 144, 208
Central & South America, 31
China, 9, 12n11, 19–21, 33n6,
  65, 66, 67n5, 68–71, 74,
  75, 77, 85n9, 86, 88, 101,
  109, 111, 131, 144n15,
  154, 158, 164, 165, 173,
  192–196, 199, 200, 207,
  211, 245, 247
Chinese
  immigrant families, 128, 169
  immigrants, 12n11, 20, 22, 69, 73,
    96, 138, 170, 189, 199,
    216, 247
  immigrant wives work, 170
  immigration, 247
  migrants, 19–21
Chinese Americans, 247n4, 249
Chinese Exclusion Act, 19–22, 247
Chiswick's hypothesis for
  understanding earnings
  differences, 38n1
Comprehensive Plan of Action (CPA),
  233, 234

D
Developing-country Asian groups, 72,
  117, 125, 127–129, 162,
  180, 223

E
Earnings, family member, 125, 127,
  129, 130
Earnings growth, vii, 1–4, 6–9, 12, 29,
  31, 36, 38, 38n1, 40–42, 45–51,
  53–62, 70–77, 71n9, 81, 84, 85,
  87–89, 92, 95, 103, 108, 114,
  117, 118, 136n6, 163, 177–180,
  177n1, 180n5, 187, 189–197,
  203, 207, 211–214, 220, 221,
  232, 236–238, 252–255
Earnings growth and human capital
  investment of immigrant, 9, 53
Eastern Europe, 4, 66, 70, 71, 73,
  173, 197
East Europeans, 96, 157n1
Education, 3, 11n9, 29–31, 30n1, 35,
  36, 38–39, 38n1, 39n2, 41, 45,
  47, 47n2, 48, 48n3, 51, 53, 54,
  56, 56n3, 58, 60, 67n4, 68, 72,
  72n10, 74, 74n14, 76, 81,
  85–92, 87n11, 98, 103, 112,
  123n7, 136, 138, 146n18, 148,
  152, 153, 162, 177–179, 180n5,
  182–184, 183n9, 183n10, 191,
  195, 204, 205, 211, 213, 214,
  218–220, 225, 226, 228,
  230–232, 234–236, 238, 249,
  252, 253, 257–262
English, 35, 37, 41, 42, 69, 70,
  95–98, 103, 107, 110n6,
  112, 132, 133, 137, 138,
  138n7, 148, 174, 192, 195,
  216–219, 217n2, 218n3,
  223, 235, 253
  investment, 95, 96

proficiency, 4, 11n9, 41, 42, 69, 70, 95–96, 98, 103, 112, 132, 138, 148, 157, 174, 189, 190, 192, 195, 195n3, 216–217, 220, 225, 227, 229, 234–238, 252
English-speaking, 41, 65, 66, 73n13, 109, 114, 217n1, 226–228, 230
English-speaking immigrants, 42
Entry earnings, vii, 3, 4, 6, 7, 9, 9n5, 10, 29–36, 38, 40–42, 45–51, 54–58, 56n3, 60, 62, 65–67, 69–78, 73n14, 77n15, 81–92, 109, 110, 110n4, 112, 114, 189–197, 212–214, 232, 236–238, 250–252, 254
European and Canadian, 9, 65, 67n5, 71, 77, 117, 121n2, 122–133, 127n14, 128n15, 137, 138, 140–143, 142n9, 145–147, 152, 153, 162, 170, 171
families, 125, 128, 128n15
immigrant, 9, 72, 77, 122, 123, 127n14, 138, 142n9, 143, 170, 195
immigrant women, 138, 142n9
women, 131, 141
European immigrants, 55–61, 73, 76, 85, 92, 98, 160, 162, 180–182, 180n5
European/Canadian group, 122, 124–132, 127n14, 140

F

Families, vii, 1, 24, 70, 81–92, 98, 117–118, 121–132, 135, 151, 163, 169–175, 177–183, 200–203, 223, 234, 247n4, 251
Family-based immigrants and refugees, 4, 252
Family business, 118, 169, 170, 173, 174, 175n6

Family income, children's contributions to, 126, 127
Family Investment Model (FIM), 117, 118, 118n3, 151–155, 159, 162, 163, 165n8, 166, 231, 232
Filipino, 11, 23, 24, 65, 69, 72, 73n14, 96, 100, 101, 109, 110n5, 122–133, 122n3, 128n15, 136–138, 138n7, 140–143, 140n8, 145–147, 152, 153, 159, 170, 171, 174, 196, 199, 248, 248n7, 249, 249n9, 263, 264
entrants, 263, 264
immigrant labor force, 264
women, 131, 138, 142n9
Filipino immigrants, 73, 111, 111n7, 184n12, 189, 199, 248, 263–264
entering, 81

G

Gentlemen's Agreement, 21, 22, 24, 248, 248n5
Group and nativity, 130, 132, 137, 140–142, 145–147

H

Hawaii, 19n1, 20, 21, 23, 247n3
High emigration rates, 107–110, 109n3, 227
Hmong, 204, 206, 211, 213, 217–219, 236–238
Human capital
investment, vii, 1–4, 9, 12, 35, 37–41, 40n2, 45–51, 55, 56n2, 58, 89, 95–104, 108, 114, 118, 151–154, 163, 180, 189–197, 200, 203, 208, 211–221, 229, 238–241, 251, 254, 255
models, 31, 38, 39
new, 4, 39–41, 58, 163, 252

Husband's self-employment, 171–174, 174n5, 175n6

I
IHCI, *see* Immigrant Human Capital Investment
IHCI Model, 4, 11, 12, 36–42, 56, 58, 74n14, 89, 90, 92, 98, 103, 118, 208, 212, 252–254
Immigrant Human Capital Investment (IHCI), 2, 4, 11, 12, 36, 37, 39–41, 56, 58, 74n14, 89, 90, 92, 98, 103, 118, 208, 212, 252, 254
Immigrants
  earnings growth, 6–8, 29, 31, 47, 48, 72–76, 74n14, 85, 252, 253
  fertility, 10n8, 144n16
  flows, 5, 19, 22, 30, 249
  high-skill transferability, 37–40, 47, 51, 252, 254
  low-skill transferability, 38–42, 47, 51, 55, 98, 103, 218, 221, 252
  networks, 184n11, 203
  permanence, 3, 254
  post-1965, 3, 7, 9, 9n5, 12, 31, 65, 67n5, 71, 77, 81, 82, 108, 187, 199, 245, 247, 252, 255
  post-reform, 1, 2, 4
  quality, 3, 9n5, 29, 41, 71, 78, 251
  self-employment, 183
  skills transfer, 40, 66
  skill transferability, 33–37, 41, 42, 78n16, 87, 110, 197, 252
Immigration Act, 4, 11, 23, 24, 248
Immigration and Naturalization Service (INS), 5n3, 83n3, 84, 84n4, 84n5, 85n9, 86, 87n10, 88, 90, 91, 178, 245, 245n1, 247, 247n2, 263
Immigration policy, 1, 4–10, 19n1, 24, 71
Indian
  immigrants, 72, 82, 129, 143, 163, 180, 180n6, 189, 190, 199, 248
  immigrant women, 143, 163
  women, 141, 159
Indochina, 208
Indochinese, 200, 201, 201n2, 203, 205, 206, 208, 211, 212, 214, 218–220, 234, 236, 237, 241, 242
Indochinese groups, 211, 218, 235
Indochinese refugees, 201, 201n3, 203, 204, 206, 208, 209, 211, 212, 216, 217n2, 218, 220, 221, 231, 234, 234n2, 235, 241, 242, 245, 247, 253
  cohort of, 200, 202, 216, 238
Initial earnings of immigrant, 1, 11, 29, 33n7, 34, 69, 70, 85, 117, 123, 179, 189
INS, *see* Immigration and Naturalization Service

J
Japan, 9, 12, 20, 21, 23, 31n5, 33, 33n6, 42, 65, 70, 73, 82, 83, 85n9, 86, 88, 101, 112–114, 144, 144n15, 173, 197, 199, 245, 248n5
Japanese immigrant families, 125n9, 128
Japanese immigrant managers, 112, 148n19

Japanese immigrants, 9, 22, 75, 112–114, 123, 123n8, 131, 140n8, 147, 148, 162, 200, 247, 254
Japanese married immigrant women, 143
Japanese migrants, 21
Japanese women, 140n8, 141, 170n3, 171

K

Korea, 9, 21, 66, 67n5, 68–71, 74, 75, 77, 82, 85n9, 86, 88, 101, 109, 111, 131, 144n15, 154, 158, 164, 165, 173, 191, 192, 196, 199, 200, 207, 208n8, 211, 245, 248n5
Korea immigrants, 81, 96, 111, 133, 143, 169–171, 173, 184n12, 189, 199, 208n8, 248
Korean immigrant businesses, 169, 173
Korean immigrant wives, 170, 173
Korean immigrant women, 133, 143
Korean women, 138, 140n8

L

Laos, 33n6, 199, 204, 211, 216, 225n3, 229, 233
Laotian immigrants, 199
Laotians, 199, 204, 206, 212, 213, 218, 223, 223n1, 225, 227–229, 231, 232, 236, 237, 241
Laotian wives, 223, 227–229
Laotian women, 225, 227, 229, 232
Low initial earnings, 2, 6, 7, 33, 47, 123, 163, 164, 166, 190, 197, 212, 213, 224, 252
Low-skill-transferability immigrants, 40, 41, 47, 51, 55, 98, 103

M

Mainland China, 66, 68, 69, 74, 75, 77, 109, 154, 158, 164, 165, 192, 195
Marriage, 121, 123, 135, 136n5, 140–142, 140n8, 141–142n9, 159
Married immigrant women, 126, 127, 129–131, 136, 137, 139–143, 145–147, 154, 157–166, 172, 173, 224–226, 228–231

N

Natives, 4, 6, 7, 10, 29–33, 37–41, 38n1, 47, 48, 50, 51, 54, 55, 58–60, 66, 67, 67n5, 69, 72–74, 73n14, 76, 77, 98, 99, 103, 110, 110n4, 111n6, 136n3, 160, 162, 175n6, 187, 189, 190, 192, 194, 195, 211–215, 217, 219–221, 236, 238, 253, 254
Non-English-speaking, 41, 65–67, 70, 95, 96, 153n3, 154, 158, 164, 165, 218

O

Occupational skills, 5, 5n4, 70, 71n8, 82, 84–86, 84n5, 84n6, 87n10, 88, 91, 92, 180–182, 182n7, 208

P

Permanence, 3, 39, 107–114, 147–148, 154, 155, 208, 221, 254
Philippines, 9, 12, 23, 24, 33n6, 67n5, 69, 71, 82, 85n9, 86, 88, 101, 111, 111n7, 144n15, 173, 199, 200, 207, 211, 217n1, 245, 248

Propensity to work of immigrant women, 166, 225

R

Refugee groups, 3, 12, 202, 216, 218–221, 223, 232
Refugees, 3–6, 5–6n4, 9, 11, 12, 24, 47n2, 65, 85, 85n7, 96, 96n3, 125n9, 199–209, 201n2, 201–202n3, 203n4, 211–221, 217n2, 223–242, 234n2, 245, 247, 252, 253, 261, 262

S

School, 21, 22, 35, 47n2, 49, 67, 67n4, 76, 86, 92, 95, 100, 101, 101n6, 103, 104, 189, 192, 194, 195, 204, 218–220, 234, 238, 241, 242, 253, 257–262
School attendance, 47n2, 100, 101, 103, 218–220, 238, 241, 242, 253
Schooling, 6, 7, 29, 34, 35, 38, 40, 47, 47–48n2, 51, 55, 61, 62, 67–69, 67n4, 67n5, 72, 75, 77, 90, 98–104, 110, 132, 135, 136, 138, 153n2, 178n4, 182, 183, 187, 189, 195, 196, 199, 203, 204, 213, 218, 219, 223, 224n2, 225, 227, 234, 236, 241, 242, 245, 249
Self-employed husbands, 170, 171, 174, 175n6
Skills
  general, 110, 135, 137, 142, 145, 146, 225, 227, 229
  transferable immigrant, 103
Skill transferability
  conceptualizations link immigrant, 34

higher immigrant, 110
low initial, 48, 89n12, 254
measures immigrant, 41
South East Asian, 200, 201, 204, 223–232
Soviet Union countries, former, 62n4, 65, 66, 109n3, 110n4

T

Taiwanese immigrants, 192, 195

V

Vietnam, 199, 204, 206, 229, 233
Vietnamese, 199, 204, 206, 211–214, 217, 220, 221, 223, 225, 225n3, 227–229, 231–236, 233n1, 238, 241, 245
  earnings, 229, 236
  wives, 225n3, 227–229

W

War brides, 125–127, 224, 226, 228, 230
Western Europe and Canada, 9, 29, 73, 99, 108, 173, 197
West European and Canadian counterparts, 162, 163
West Europeans, 4, 30, 31, 35, 62, 70, 72, 73, 96, 100, 101, 108, 114, 153n3, 157, 159, 162, 163, 192, 195, 200, 206, 219, 223–232, 225n3, 231n7, 234, 251, 253
Work, propensity to, 118, 131, 136, 136n5, 138, 144, 146, 148, 152–155, 159–164, 166, 173, 174, 174n5, 175n6, 225–227, 231, 232

CPSIA information can be obtained
at www.ICGtesting.com
Printed in the USA
LVHW080636150622
721320LV00005B/137